Palgrave Studies in Digital Business & Enabling Technologies

Series Editors
Theo Lynn
Irish Centre for Cloud Computing (IC4)
Dublin City University
Dublin, Ireland

John G. Mooney
Graziadio Business School
Pepperdine University
Malibu, CA, USA

This multi-disciplinary series will provide a comprehensive and coherent account of cloud computing, social media, mobile, big data, and other enabling technologies that are transforming how society operates and how people interact with each other. Each publication in the series will focus on a discrete but critical topic within business and computer science, covering existing research alongside cutting edge ideas. Volumes will be written by field experts on topics such as cloud migration, measuring the business value of the cloud, trust and data protection, FinTech, and the Internet of Things. Each book has global reach and is relevant to faculty, researchers and students in digital business and computer science with an interest in the decisions and enabling technologies shaping society.

More information about this series at
http://www.palgrave.com/gp/series/16004

Theo Lynn · John G. Mooney
Pierangelo Rosati · Mark Cummins
Editors

Disrupting Finance

FinTech and Strategy in the 21st Century

Editors
Theo Lynn
DCU Business School
Dublin City University
Dublin, Ireland

Pierangelo Rosati
DCU Business School
Dublin City University
Dublin, Ireland

John G. Mooney
Graziadio Business School
Pepperdine University
Malibu, CA, USA

Mark Cummins
DCU Business School
Dublin City University
Dublin, Ireland

Palgrave Studies in Digital Business & Enabling Technologies
ISBN 978-3-030-02329-4   ISBN 978-3-030-02330-0   (eBook)
https://doi.org/10.1007/978-3-030-02330-0

Library of Congress Control Number: 2018957678

PREFACE

This second volume in *Palgrave Studies in Digital Business & Enabling Technologies* further contributes to multidisciplinary research on digital business and enabling technologies in Europe by exploring the evolving domain of the next generation of financial technologies or "FinTech". The concept of evolution is important in this context as FinTech is not a new concept. Since the 1950s, each decade has witnessed a new technology that has transformed how financial services operate and how we interact with them. Credit card processing, ATMs, electronic stock trading, e-commerce are just some of the myriad of technologies that we take for granted. Today, we are seeing the advent of a new generation of FinTech built on near-ubiquitous access to the Internet through mobile and cloud computing, machine learning, artificial intelligence and blockchain. These technologies are resulting in significant disruptive changes to the financial services sector, not least opening up the sector to increased competition and empowering customers in ways unthinkable just a decade ago.

While practice might view FinTech as a co-evolution and convergence of finance and technology, one could be mistaken in thinking that for finance researchers it is business as usual. Finance research is concerned with risk and return framed by established theories such as asset pricing theory, modern portfolio theory and the efficient market hypothesis, albeit with emerging challenges from the set of theories underpinning behavioural finance. Yet, it is clear that FinTech is changing, as Drucker (1994) might put it, the 'theory of the business' or 'mental models' upon which the financial sector is based. In the same way that established financial services firms, banks, and insurance companies are being forced to rethink their role and activities in the market, finance researchers need to reflect on the impact of FinTech innovation on finance research. What are the implications of FinTech innovation for finance literatures? As FinTech represents greater convergence of finance and technology, is greater collaboration between the finance and technology research domains required to ensure greater relevance and market impact? Does FinTech represent a new discipline in itself? While this book does not seek to address these questions, it has value to university educators and researchers, industry practitioners, and policymakers as an entrée into the wider FinTech ecosystem and some of the extant, although early stage, research being undertaken in this space.

Addressing the call for inter-disciplinarity, contributors have been drawn from an international group of scholars in finance, law, computer science and management. "Disrupting Finance" presents a variety of perspectives on how technologies are making us rethink lending, regulation and compliance, risk management, insurance, stock trading, payments, and money in the fourth industrial age. FinTech is changing how individuals, projects and businesses access finance and from whom. Chapters 1 and 2 discuss crowdfunding and online peer-to-peer (P2P) lending, a form of crowdfunding that bypasses conventional intermediaries, processes and requirements to connect borrowers and lenders. Information asymmetry is a key issue in lending that can result in moral hazard or adverse selection. Chapter 2 explores this issue specifically discussing some of the mechanisms being used by online P2P lending platforms to reduce this risk. The theme of risk management is continued in Chapter 3 where the role of machine learning and artificial intelligence is discussed in the context of the assessment and management of credit risk, market risk, operational risk, and compliance.

Chapter 4 presents a thematic analysis of extant literature on the somewhat controversial area of high-frequency trading and discusses key themes in extant literature including the impact of high frequency trading (HFT) on market liquidity, trading strategies and speed, implications for market structure changes, and the relationship between the "scriptability" of corporate disclosure and short-term information advantage. Asymmetric information and the use of new data science techniques is a common theme in many of the chapters. Chapter 5 deals with emerging uses cases in InsurTech and specifically how large and continuous datasets are transforming general insurance markets and their business processes, modifying policyholder behaviour, and streamlining claims management. The authors illustrate how machine learning, artificial intelligence and blockchain are creating and helping to capture new value in the insurance market.

A common theme in each segment of financial service sector are the barriers to entry created by regulation. Indeed, the lending, insurance, and stock markets are all characterised by regulatory requirements that are complex to understand and costly to implement for incumbents and new entrants alike. With over US$100 billion spent by banks on regulatory compliance in 2016 alone, RegTech solutions represents a significant market opportunity in itself by (a) identifying the impacts of regulatory provisions on business models, products and services, functional activities, policies, operational procedures and controls; (b) enabling compliant business systems and data; (c) helping control and manage regulatory, financial and non-financial risks; and (d) performing regulatory compliance reporting. Chapter 6 explores the drivers of RegTech adoption and the risks and challenges inherent in this adoption. It presents a timely focus on the lack of standardisation and interoperability in RegTech data and systems and the need for open standards and semantic technologies in order to avoid a digital Tower of Babel in the financial sector.

Chapters 7 and 8 focus on the future of payment and money. The European Union required its member states to implement the new Payment Service Directive (PSDII) in January 2018. This directive has the potential to drastically reimagine the relationships between consumers and their banks and the structure of the banking and payments sector. Driven by the Internet and mobile banking and the need for more efficient and effective support for cross-border payment services, PSDII seeks to level the competitive playing field by reducing the various exemptions from payment services regulation and to permit two new innovative arrangements: "account information service providers" and "payment initiation service providers". Chapter 7 presents the background and detail of PSDII and the implications for banks, credit card issuers, merchant acquirers and new FinTech operations, not least technology firms such as Apple, Google, PayPal etc. While Chapter 7 reimagines the role of banks in the payment sector, Chapter 8 discusses the reconceptualisation of money in the digital age. This chapter explores the characteristics of money and the affordances of digital money which make it something very different—frictionless, anonymous, transparent, non-denominated and dataful. Furthermore, the authors discuss the concept of money as opportunities for social encounters in transactions with very real social impacts.

The final two chapters focus on the related topic of cryptocurrencies and blockchain. Following on from this discussion on the future of payment and money, Chapter 9 focuses on cryptocurrencies as two distinct flavours of digital token—native coins and crypto tokens. While native coins are well known as a new form of digital money such as bitcoin, crypto tokens are less well known. They represent a form of "digital vouchers" that allow the token holders to get access to almost any type of service and assets from monetary rewards, or commodities to loyalty points to even other cryptocurrencies. As well as discussing the differences between these token based models, Chapter 8 explores the emergent start-up token funding model of Initial Coin Offerings (ICOs), which allows entrepreneurs to bypass the traditional capital market by issuing crypto tokens out of thin air. Blockchain, or distributed ledger technology, is one of the most hyped technologies in recent years and no FinTech book would be complete without a wider discussion on it. Chapter 10 discusses the current challenges and opportunities that blockchain poses for financial services firms and its potential impact on four main financial activities: (1) payments and remittance, (2) credit and lending, (3) trading and settlement, and (4) compliance.

The ten chapters in "Disrupting Finance" are by no means exhaustive nor were they intended to be. Rather the collection of topics in this book were collated to be a primer and signpost for FinTech. The financial services sector is a dataful one—it comprises data and generates data. It is unsurprising therefore that technologies that enable the exchange, validation, and analysis of this data faster and in more complex ways dominate the FinTech discourse today. Blockchain, deep learning and artificial intelligence are not only challenging how we conceive financial services but introduce new avenues for research not just in finance and computer science but in ethics, sociology and law, to name but a few.

Dublin, Ireland   Theo Lynn
Malibu, USA   John G. Mooney
Dublin, Ireland   Pierangelo Rosati
Dublin, Ireland   Mark Cummins

ACKNOWLEDGEMENTS

This book was supported by the Irish Centre for Cloud Computing and Commerce, an Irish National Technology Centre funded by Enterprise Ireland and the Irish Industrial Development Authority.

CONTENTS

10   Blockchain Beyond Cryptocurrencies

Pierangelo Rosati and Tilen Čuk

Notes on Contributors

Saqib Aziz is Assistant Professor of Finance in Rennes School of Business in France. He obtained his Ph.D. in Finance from Rennes University, France. His primary research interests are in stability of financial institutions, an analysis that cuts across investigating various types of risk and its genesis in mergers and acquisitions led growth strategies, safety-net distortions and national culture, financial regulations, and AI driven risk management.

Alan Brener is a Teaching Fellow at University College London's Law Faculty. Alan is Deputy Director of the Centre for Ethics and Law at University College London and a Council Member of The Chartered Institute of Bankers in Scotland. Alan worked for Santander UK for some ten years and was responsible, at different times, for the compliance and retail legal departments and regulatory policy. Prior to this he worked in senior positions at other banks and as a regulator and in central government.

Tom Butler is Professor at the Department of Accounting Finance and Information Systems University College Cork. He has published widely in the IS field's leading journals and conferences, with over 200 publications. He has been awarded over €8m in research funding and has several technology innovations. A global thought leader in the emerging field of RegTech, he is co-founder of SemanticRule, a RegTech spinout from UCC.

Dominic Cortis is a lecturer with Faculty of Economics, Management and Accountancy at The University of Malta. He is an associate actuary and is a non-executive director of a non-life insurer. Dominic has published academic papers focusing on the challenges faced in the insurance and sports betting industries.

Tilen Čuk is a Ph.D. candidate at the Perelman Centre for legal philosophy (Université libre de Bruxelles). He has previously studied at the Ecole normale supérieure (Cachan), Paris I Sorbonne and Oxford University (MJur). His research activities focus between law and economics, with a particular interest in artificial intelligence, blockchain and algorithmic and high-frequency trading. Tilen is also a co-founder of the knowledge hub FinTech Policy EU.

Mark Cummins is Professor of Finance at the DCU Business School. He holds a Ph.D. in Quantitative Finance with specialism in the application of integral transforms and the fast Fourier transform (FFT) for derivatives valuation and risk management. Mark has research interests in an array of areas, including computational finance, model risk management, and the emerging areas of FinTech and InsurTech. He has published in leading international finance journals, as well as leading international field journals. He is co-editor of the "Springer Proceedings in Mathematics" title Topics in *Numerical Methods for Finance* and is Associate Editor with *Finance Research Letters*.

Jeremy Debattista is a Research Fellow within the ADAPT Centre at Trinity College Dublin. He currently holds a grant from the Irish Research Council. Jeremy's research area focuses on challenges related to data quality, Linked Data, and applied AI.

Johann Debono is a B.Com. (Hons.) Insurance graduate from the University of Malta and works as a senior broker with one of the leading insurance brokers in Malta. He is in the final stages of completing a Masters in Risk Management through Birmingham City University with a research focusing on cyber risk and the use of cyber liability insurance as risk mitigation tool.

Francesca Di Pietro is Assistant Professor in Business Strategy at Trinity Business School, Trinity College Dublin, the University of Dublin. She was a Postdoctoral research fellow at LUISS Guido Carli University in Rome, Italy. She has been visiting scholar at Aalto University in Finland,

HEC School of Management in Paris, and Cass Business School in London. She holds a Ph.D. in Management and Business Administration from University G. d'Annunzio, Pescara. Her main research interests are in the areas of entrepreneurship and entrepreneurial finance.

Michael Dowling is Associate Professor of Finance in Rennes School of Business in France. He obtained his Ph.D. in behavioural finance from Trinity College Dublin, Ireland. Michael is Director of the AI-driven Business research group in Rennes, and through this explores how organisations can benefit from, and adapt to, the arrival of AI in the organisation. He is also editor of the leading behavioural finance journal: *Journal of Behavioral and Experimental Finance.*

Mark Farrell is a Fellow of the Institute and Faculty of Actuaries (FIA) and Senior Lecturer at Queen's University Belfast. He is Programme Director of Actuarial Science and Risk Management at Queen's Management School. He has a Ph.D. in Finance and is a Fulbright Scholar. Prior to moving into academia, he spent 10 years working as a consulting actuary in London, Toronto, Belfast and Dublin.

Jennifer Ferreira teaches and conducts research in the field of human–computer interaction. Her publications span three broad themes: the digital economy, agile development and user experience design, and user interface design and evaluation. She applies qualitative methods in her work, with an emphasis on ethnographically-informed and participatory design approaches. Her interest in financial services and digital money began as part of the 3DaRoC project, which explored the ways digital connectivity shapes peer-to-peer relationships in alternative financial services.

Theo Lynn is Professor of Digital Business at Dublin City University and is the Principal Investigator (PI) of the Irish Centre for Cloud Computing and Commerce, an Enterprise Ireland/IDA-funded Cloud Computing Technology Centre. Professor Lynn specialises in the role of digital technologies in transforming business processes.

Ciarán Mac an Bhaird is Assistant Professor of Economics and Finance at Fiontar (Enterprise), Dublin City University and founder of USTART, the DCU student Start-up accelerator. His research is focused on Entrepreneurial Finance, specifically capital structure, financial management, resourcing nascent firms, and alternative sources of finance.

Eleonora Monaco is Assistant Professor of Accounting and Finance at Católica Porto Business School (Portugal) where she teaches courses in financial accounting, financial statements analysis and accounting quality. She holds a Ph.D. in accounting from University of Chieti (Italy) and a postdoc spent at Capital Markets Cooperative Research Centre in Sydney. During her career, she has been appointed Visiting Fellow both at the University of Edinburgh Business School and at Queensland University of Technology in Australia. Her primary research areas include capital markets-based accounting research, M&A and earnings management.

John G. Mooney is Associate Professor of Information Systems and Technology Management and Academic Director of the Executive Doctorate in Business Administration at the Pepperdine Graziadio Business School. Dr. Mooney previously served as Executive Director of the Institute for Entertainment, Media and Culture from 2015–2018, Associate Dean for Academic Affairs and Online Programs from 2014–2015; Department Chair for Strategy, Entrepreneurship, Information Systems and Decision Sciences from 2012–2014; and as Associate Dean for Academic Programs from 2005–2010. He was Conference co-Chair for the 2016 37th International Conference on Information Systems that was held in Dublin, Ireland. Dr. Mooney holds a B.S. in Computer Science and a Master of Management Science both from University College Dublin, and a Ph.D. in Information Systems from UC Irvine. He was a Visiting Scholar at the MIT Sloan Center for Information Systems Research from 2010–2011 and continued his affiliation as Research Associate until 2016. His current research interests include management of digital innovation (i.e. IT-enabled business innovation) and business executive responsibilities for managing digital platforms and information resources.

Leona O'Brien, LL.M., B.C.L., BBus is Senior Legal Researcher at the Department of Accounting Finance and Information Systems. Her research interests are Semantic Technologies, Financial Regulation and RegTech/FinTech. Leona has attracted over €440,000 research funding as co-PI and is co-founder of SemanticRule, a RegTech spinout from UCC.

Mark Perry is an interdisciplinary user studies researcher. His work involves evaluating the ways that digital technology is practically used by people in support of interaction design, with research covering the areas of mobile and ubiquitous technology, digital and mobile money, and emerging financial services (FinTech). He has been funded to work on a number of research projects exploring user experience, stakeholder perspectives, and systems design around digital connectivity and peer-to-peer relationships in financial services.

Pierangelo Rosati is Assistant Professor in Business Analytics at DCU Business School. He previously worked as Post-Doctoral Researcher of the Irish Centre for Cloud Computing and Commerce (IC4). Dr. Rosati holds a Ph.D. in Accounting and Finance from the University of Chieti-Pescara (Italy) and an M.Sc. in Management and Business Administration from the University of Bologna. He was appointed Visiting Professor at the Universidad de las Américas Puebla and at Católica Porto Business School, and visiting Ph.D. Student at the Capital Markets Cooperative Research Center (CMCRC) in Sydney. Dr. Rosati has been working on research projects on FinTech, Blockchain, cloud computing, data analytics, business value of IT, and cyber security.

Paolo Tasca is a Digital Economist specialising in P2P financial systems. An advisor on blockchain technologies for international organisations such as the EU Parliament and the United Nations, Paolo is founder and Executive Director of the Centre for Blockchain Technologies at University College London (UCL CBT). Previously, he was Lead Economist on digital currencies and P2P financial systems at the Deutsche Bundesbank in Frankfurt.

LIST OF FIGURES

LIST OF TABLES

CHAPTER 1

Deciphering Crowdfunding

Francesca Di Pietro

Abstract Crowdfunding is one of the funding sources that entrepreneurs are increasingly being exposed to above and beyond venture capital and angel funding. After introducing the most popular crowdfunding types, this chapter proceeds to present and compare the evolution of the phenomenon across three macro regions: Europe, USA, and Asia-Pacific. Furthermore, the chapter offers an overview of the state-of-the-art of the crowdfunding literature, highlighting creators' and funders' incentives and disincentives for starting or engaging in crowdfunding projects, the characteristics of a successful campaign, and the contextual factors that explain the evolution of the phenomenon across countries. The chapter closes providing suggestions for future work in this area.

Keywords Crowdfunding · Geographic markets · Crowdfunding platforms · Successful crowdfunding campaign characteristics

F. Di Pietro (✉)
Trinity Business School, Trinity College Dublin,
The University of Dublin, Dublin, Ireland
e-mail: francesca.dipietro@tcd.ie

© The Author(s) 2019
T. Lynn et al. (eds.), *Disrupting Finance*, Palgrave Studies
in Digital Business & Enabling Technologies,
https://doi.org/10.1007/978-3-030-02330-0_1

1

1.1   The Crowdfunding Phenomenon: An Overview

Crowdfunding can be defined as "an open call, essentially through the Internet, for the provision of financial resources in order to support initiatives for specific purposes" (Belleflamme et al. 2014, p. 588). Mollick (2014), narrowing the definition in an entrepreneurial context, defines crowdfunding as the efforts by entrepreneurial individuals and groups–cultural, social, and for-profit–to fund their ventures by drawing on small contributions from a relatively large number of individuals using the Internet. Crowdfunding draws inspiration from the concept of microfinance (Morduch 1999) and crowdsourcing. It represents a unique form of fundraising where capital seekers (project proponents) are linked with capital givers (investors) through a crowdfunding intermediary (platform) (Haas et al. 2014).

During the past five years different forms of crowdfunding have emerged. Based on the risk of funding for investors, we can distinguish between investment and non-investment crowdfunding models. Within these two groups, according to the right of crowdfunders in the projects' outcome, crowdfunding platforms can further be categorised as follows:

Investment models:

- Lending-based crowdfunding: funds are paid back and funders have the right to receive a contractually agreed interest payment. This model is further categorised into two major submodels–(1) peer-to-peer lending (P2P) which is characterised by direct interaction between the two parties (see Chapter 2), and (2) social lending, usually used for entrepreneurial projects at local level.
- Equity-based crowdfunding: funds are provided in exchange for company's shares. Investors have the right to receive returns on investments if the company performs well.

Non-investment models:

- Reward-based crowdfunding: funds are provided in exchange for non-monetary benefits. Common benefits include a small gift (reward) or a reservation for a product which is still under production (pre-order).
- Donation-based crowdfunding: funds are provided for philanthropic or sponsorship reasons with no expectation of remuneration.

Crowdfunding platforms can also be categorised at a macro level. Generalist platforms enable crowdfunding for any area of interest while vertical (or thematic) platforms focus on crowdfunding for projects within a specific field or sector.

Lastly, crowdfunding platforms can be distinguished based on the different funding mechanisms adopted. Platforms can regulate the pledge levels, the minimum investment amounts, and decide whether to adopt an "all or nothing" or "keep it all" funding principle (Gerber et al. 2012; Mollick 2014; Cumming et al. 2015). The "all or nothing" funding approach allows project proponents to receive funding only if the campaign achieves 100% of the target (Belleflamme et al. 2010; Cumming et al. 2015; Haas et al. 2014). If the target amount is not met, investors receive their money back. On the other hand, the "keep it all" funding approach allows project proponents to receive any collected amount (Gerber et al. 2012).

1.1.1   The European Market

Since its inception, the crowdfunding market in Europe has experienced exponential regional and country level growth (Cambridge Centre for Alternative Finance 2018). Overall the market volume increased by 41% annually, from €5.431 billion in 2015 to €7.671 billion in 2016. This growth was driven by dramatic growth in the UK market (accounting for 73% of the entire European market alone), and the fast expansion of alternative finance markets in smaller European countries like the Nordics, the Iberian Peninsula, and the Baltics. The second largest European market is France, accounting for 22% of the European crowdfunding market in 2016, followed by the German market (15.6%).

In 2016, P2P lending was the most popular crowdfunding investment model in Europe. P2P consumer and business lending accounted for 33.8% (€696.81 million) and 17% (€349.96 million), respectively of the total crowdfunding market. In contrast, equity-based and reward-based crowdfunding experienced a decline in 2016, each accounting for approximately 10% of the market. Donation-based platforms grew modestly in 2016 however only represent 1.6% of the market share.

The further development of the crowdfunding market is challenged by risk factors perceived by prospective investors. In the investment models and reward-based model, for instance, two main risks highlighted by potential investors are (1) the risk of fraud, e.g. the possibility that the product or project announced during the campaign may be false or non-existent and thus the fundraiser could attempt to use the funds collected from the backers for other (personal) purposes; and (2) the risk of platform collapse due to malpractice by fundraisers (Cambridge Centre for Alternative Finance 2018). Other risks, such as the loan default or late repayments to investors, in the case of lending-based crowdfunding, as well as changes in local and/or EU regulation were perceived as less critical for investors (Cambridge Centre for Alternative Finance 2018). It is noteworthy that in the UK, unlike the general EU outlook, cybersecurity was viewed as a major risk factor and the impact of Brexit was not a major concern (Cambridge Centre for Alternative Finance 2017c).

1.1.2   The US Market

In the United States, one of the world's largest and most innovative countries for alternative finance, the total volume raised via crowdfunding in 2016 was US$34.5 billion. Regulatory interventions, such as the signing of JOBS ACT, in April 2012, have significantly contributed to the development of the crowdfunding market. Specifically, the Tittle III of the JOBS ACT brought big changes within the entrepreneurial finance landscape, alleviating the burden of entrepreneurs' in raising finance, allowing fundraising from large number of investors through approved intermediary portals.

Like in the European market, P2P lending is the dominant crowdfunding model in the United States, accounting for 61% of the market in 2016, followed by the equity-based model which delivered US$549 million to approximately 637 businesses and reward-based crowdfunding (US$551 million) (Cambridge Centre for Alternative Finance 2017b). Data also reveals a slight drop in equity-based crowdfunding compared to 2015, which seems to be related to regulatory ambiguity, also a feature of the EU market. Among the non-investment models, donation-based crowdfunding expanded a great deal in 2016 with US$224 million raised, while reward-based crowdfunding suffered a decline from 2015 to 2016, but still continues to attract significant funds from crowds (Cambridge Centre for Alternative Finance 2017b).

According to investors' risk perceptions, in the United States the risk of fraud involving high-profile campaigns and business failure rate were viewed as the greatest risks to the crowdfunding industry (Cambridge Centre for Alternative Finance 2017b).

1.1.3   The Asia-Pacific Market

Another area of interest for the development of the crowdfunding market is the Asia-Pacific region with the key regional markets being China, Oceania, and South Asia. China is the market leader accounting for 99.2% of the total Asia Pacific crowdfunding market. In 2016, the total volume of transactions in China was US$245.38 billion, an annual growth of 136% compared to 2015. Like the EU and the US, P2P lending represents the main segment of interest with 56% of the total market in China (Cambridge Centre for Alternative Finance 2017a).

Outside of mainland China, Australia has shown high growth rates, reaching US$609 million in total volume in 2016, followed by Japan (US$400 million), and South Korea (US$376 million). P2P lending in the Australian market accounts for about 60% of the total market (Cambridge Centre for Alternative Finance 2017a). In South Asia, the prominent country active in the alternative finance industry is India, with a total of $124 million collected in 2016. From 2013 to 2016 Indian activity was mainly concentrated in P2P lending, accounting for about 60% of the market. Equity-based and donation-based crowdfunding each accounted for around 15% of the total crowdfunding market volume, representing US$32.3 million of funds raised (Cambridge Centre for Alternative Finance 2017a). A distinctive aspect of the Indian economy is the lack of access to bank credit for the majority of the population. In fact, only 10% of the 1 billion inhabitants have access to it. The growth of crowdfunding, allowing people to connect via the Internet and access financial services that are not available elsewhere, could benefit a huge portion of the population.

Overall, the latest statistics suggest that over the course of the past three years in the three main macro-regions, Europe, USA, and Asia-Pacific, the alternative finance market doubled its volumes and is continuing to growth impressively. P2P crowdfunding and reward-based crowdfunding are consistently the most popular models in all markets.

1.2   CROWDFUNDING STATE-OF-THE-ART

Although crowdfunding is a relatively new method of funding for start-ups and small ventures, it has become an increasingly relevant means of alternative financing. Researchers and scholars have started to investigate this phenomenon in the attempt to construct a theoretical model that could fully explain the dynamics behind crowdfunding, focusing on two main aspects: (1) the incentives and disincentives for starting or taking part in crowdfunding projects (see, e.g., Gerber et al. 2012; Agrawal et al. 2014; Allison et al. 2015) and (2) factors associated with successful campaigns (see, e.g., Agrawal et al. 2011; Mollick 2014; Colombo et al. 2015; Giudici et al. 2018).

Investigating initiators' incentives for starting a crowdfunding project, research shows that fundraisers consider crowdfunding as an opportunity (1) to finance their company at a lower cost of capital; (2) to receive public attention; and (3) obtain feedback on the product or service offered (Gerber et al. 2012; Agrawal et al. 2014; Belleflamme et al. 2014). Crowdfunding platforms are easily accessible and thereby represent an opportunity for entrepreneurs to test the marketability of the project and receive suggestions. Nevertheless, embracing crowdfunding entails a great deal of public exposure and information disclosure (Agrawal et al. 2014). If creators are unable to collect the necessary funds from the crowd they will face the threat of not only reducing their chances of receiving future investments but also that others may steal their ideas.

Funders, on the other hand, finance crowdfunding campaigns to support an innovative idea, to help others to realise their dreams, to gain early access to new products, and to be part of a community (Zhang 2012; Agrawal et al. 2014). The exchange of resources, followed by continuous interactions among members (both supporters and creators) of these platforms, generates a sense of belonging to a community where individuals share similar views and beliefs. Nonetheless, funders face some disincentives in engaging in crowdfunding projects. Early-stage companies that generally approach crowdfunding are inherently risky and often funders and creators are overoptimistic about projects outcomes (Mollick 2014). Moreover, crowdfunding is a fertile ground for fraudulent behaviours. Creators could provide false information to promote their company and information asymmetries are very high (Agrawal et al. 2014). Lastly, in crowdfunding, virtual meetings replace real-life encounters, making it more challenging for the crowd to understand what businesses and what intermediary can be trusted (Schwienbacher and Larralde 2010).

1.2.1   Investment Models

Although it represents the greatest share of the crowdfunding market, there is a relatively small literature on lending crowdfunding. Lending crowdfunding is best suited for ventures that have built a viable product and generates some initial revenue, demonstrating early traction (Paschen 2017). Scholars have analysed the role of networks within peer-to-peer lending crowds and their effect on crowdfunding campaigns' performance. Network relations provide larger proportions of loans, lending four times more than strangers. Investors with preexisting network ties also respond to loan requests on average 59.5% sooner than strangers (Horvát et al. 2015). These results are in line with extant research on other crowdfunding models highlighting the importance of relations and network in online fundraising success (Agrawal et al. 2011; Mollick 2014; Colombo et al. 2015; Butticè et al. 2017).

Scholars have further investigated the lending behaviour in P2P crowdfunding platforms. Cummins et al. (2018), comparing lending practices of non-banking institutional investors and retail investors, found that institutional investors generally outperform retail investors, achieving higher returns upon repayments and having a lower likelihood of loan default than retail investors. Along the same lines, Kgoroeadira et al. (2018), investigated whether P2P small business lending has different characteristics than traditional small business lending. Unlike traditional lenders, P2P online lenders, in deciding whether or not to lend money to businesses, focus more on entrepreneurs' personal characteristics—e.g. person's credit score, employment, picture, etc.— than business characteristics. This suggests that entrepreneurs should approach online markets, tailoring their pitch as personal rather than focusing on firm characteristics, since the latter are the main determinants of securing funding.

Similarly, while equity crowdfunding has reached significant investment volumes, the number of research studies on the area is relatively small. One of the first studies tackling the equity crowdfunding phenomenon was conducted by Ahlers et al. (2015) examining the impact of project quality (human capital, social capital, and intellectual capital) and perceived level of uncertainty on the success of a campaign. The authors found that while entrepreneur's human capital was relevant in attracting a higher number of investors and capital, social, and intellectual capital did not appear to be key success factors. They also highlighted the importance of providing detailed information about the company—e.g. financial roadmaps, risk factors, etc.—to prospective backers to reduce information asymmetry. Nevin et al. (2017), focusing on the role of social media activities in equity-crowdfunding campaign success, show that being active on social media, engaging with the crowd, and understanding social media selectivity—using different social media according to the target audience—positively impact the outcome of a crowdfunding campaign. Lynn et al. (2017) provide insights on the crowdfunding network on Twitter—comprising multiple subcommunities, hubs, and influencers–illustrating the geographical concentration of crowdfunding in specific areas or communities and highlighting the importance of the social media use during crowdfunding campaigns.

Looking at the post-equity funding performance of crowd-backed start-ups in the UK, Signori and Vismara (2018) found that 18% of them were not active anymore, whereas 35% raised further funding from either traditional investors (9%) or follow-on crowdfunding offering (25%). Lastly, taking a qualitative perspective, the study by Di Pietro et al. (2018c) illustrates how entrepreneurs leveraging investor networks generated in the course of equity-based crowdfunding campaigns, contributes to the success of start-up firms. Crowd investors can provide firm founders with knowledge related to the product, strategy, and market as well as network ties with industry players and other stakeholders.

1.2.2   Non-investment Models

Agrawal et al. (2011) conducted one of the first studies investigating how online reward-based crowdfunding platforms reduce investors' costs related to early-stage financing (e.g. collecting initial information, monitoring, providing inputs, etc.) and eliminate most of the distance-related economic frictions. Their study shows that Family and Friends (F&F) investors are mostly local and invest in the project in the early phases of the funding process, whereas non-F&F investors are more geographically disperse and willing to fund as the capital raised increases. Mollick (2014), taking a broader perspective, focused on the role of the founders themselves in determining the success (or failure) of their online reward-based campaigns. The perceived quality of the underlying project, which can be signalled, for example, by including videos, providing frequent updates, and avoiding spelling errors, together with the founders' social network size, increase the chances of success of a crowdfunding campaign. The effects of geography on the success of the project were also considered, in terms of proximity to funders, supporting Agrawal et al. (2011) findings.

Colombo et al. (2015), investigating reward-based crowdfunding from a social capital perspective, show that during the first stages of the campaign, when the level of uncertainty concerning the proposed projects is high, the founders' relationships within the crowdfunding community are necessary to spread information and attract early backers. The relationships and social contacts developed among founders and backers within the same online platform—*internal social capital*—appears to be crucial in attracting both early-capital and early-backers, which are strong predictors of a campaign's success. Internal social capital, reducing the information asymmetry between founders and backers, triggers imitating behaviours of other investors that feel more confident in endorsing the proposed project. Along this line, Butticè et al. (2017) showed that serial projects' proponents in reward-based crowdfunding platforms were able to build internal social capital by launching several successful campaigns, increasing their chances of succeeding also in their subsequent campaigns. By supporting the same project throughout its whole creative process, backers become part of a virtual community sharing common views and goals and this creates an emotional connection with other members of the group and with the entrepreneur as well.

The importance of social capital in reward-based crowdfunding and, in general, the characteristics of the geographic area where project's proponent reside were also examined by Giudici et al. (2018). Their study demonstrates that *local altruism*—the level of altruism shared by people living in the founder's city—represents a key success factor for crowdfunding projects and this effect is strengthened by the entrepreneur's personal social networks, supporting the findings of Mollick (2014), and *local relations* among residents. More recently, Di Pietro et al. (2018a), looking at the characteristics of the geographic area where investors reside, suggest that *local religiosity* can play a significant role in enhancing the fundraising of cross-regional crowdfunding projects.

On the topic of donation-based crowdfunding, scholars have focused their attention mainly on the determinants of funding behaviours (Gleasure and Feller 2016), considering the benefits that donors achieve through donations. In some cases, financial benefits are gained in the form of tax deductibility as suggested by Meer (2014), but more frequently social benefits represent the main reward for donors. From this perspective, the way in which projects are described and the anonymity of users influence the propensity and the amount of donations

(Smith et al. 2013; Burtch et al. 2015). Di Pietro et al. (2018b), analysing crowdfunding donations collected in Italy by Mary's Meals Charity over a 15-month period, illustrate that (lower) *digital divide*, i.e. access to infrastructure such as broadband connection, and user's *digital literacy*, i.e. habitual use of Internet, enhance funder's propensity to use a crowdfunding platform for donation purposes. The findings also support the hypothesis that the use of social media, in particular Facebook sharing, positively influences donations.

Overall, research in both investment and non-investment crowdfunding models illustrate that (1) funding is not geographically constrained although geographical proximity matters since most of the capital flows to the same regions (e.g. Agrawal et al. 2011; Lynn et al. 2017; Cambridge Centre for Alternative Finance 2018); (2) funding propensity grows with collected capital, highlighting the importance of friends and family as well as entrepreneurs' internal social capital as they invest early in the funding cycle (e.g. Agrawal et al. 2011; Mollick 2014; Colombo et al. 2015; Butticè et al. 2017); and that (3) funding is influenced by the characteristics of the local area in which entrepreneurs and investors live (e.g. Di Pietro et al. 2018a; Giudici et al. 2018).

1.3   New Research Trends: The Language of Crowdfunding

Recent studies are exploring the relevance of linguistic style in entrepreneurial finance, a relatively nascent research area. It is widely accepted that the way entrepreneurs articulate their business is vital for fundraising and legitimacy purposes. Prior research highlights storytelling as a means for entrepreneurs to establish venture legitimacy and gain stakeholder support. Specifically, Gorbatai and Nelson (2015) focus on the linguistic features of crowdfunding platforms to test the effect of language on the outcome of alternative financing campaigns and the relationship between linguistic content and gender. The authors suggest that the three types of language most likely to be successful in crowdfunding campaigns are positive, vivid, and inclusive, while business language would be less rewarding for the outcome of a campaign. Davis et al. (2017) support the idea that product originality and passion displayed by entrepreneurs in their pitches strengthen entrepreneurs' ability to collect funding. Similarly, Parhankangas and Renko (2017) compare the linguistic style of commercial and social entrepreneurs' pitches, finding that whereas language style is of little importance to commercial entrepreneurs, social entrepreneurs need to pay more attention to *how* they deliver their pitches: the use of concrete language, precise terminology, and an interactive style, are fundamental for the success of a social crowdfunding campaign. The perception that an entrepreneur conveys to potential investors is of fundamental importance since the financing pitch is often made before or during the development stage, that is, before the audience can actually see the finished product. Future research can build on these very preliminary findings and explore the role of communication and storytelling in crowdfunding in much greater depth.

Another important question to address in future research is related to factors that help us understand cross-country and cross-regional differences in the development of crowdfunding. In fact, despite a significant variation in entrepreneurial financing activity across countries (Reynolds 2011; Cambridge Centre for Alternative Finance 2018), the theoretical explanations of the antecedents and processes are limited (Baker et al. 2005). Drawing from cognitive psychology theories, recently Di Pietro et al. (2018d) show that linguistic structures may help in explaining differences in entrepreneurial finance market dynamics across nations and cultures. More interdisciplinary research is needed to discover underlying mechanisms that will help us to understand international differences of the crowdfunding phenomenon.

REFERENCES

Agrawal, A., Catalini, C., & Goldfarb, A. (2011). *Friends, family, and the flat world*. Toronto: The Geography of Crowdfunding.

Agrawal, A., Catalini, C., & Goldfarb, A. (2014). Some simple economics of crowdfunding. *Innovation Policy and the Economy, 14*(1), 63–97.

Ahlers, G. K. C., Cumming, D., Günther, C., & Schweizer, D. (2015). Signaling in equity crowdfunding. *Entrepreneurship Theory and Practice, 39*(4), 955–980.

Allison, T. H., Blakley, C., Davis, J., Short, C., & Webb, J. W. (2015). Crowdfunding in a prosocial microlending environment: Examining the role of intrinsic versus extrinsic cues. *Entrepreneurship Theory and Practice, 39*(1), 53–73.

Baker, T., Gedajlovic, E., & Lubatkin, M. (2005). A framework for comparing entrepreneurship processes across nations. *Journal of International Business Studies, 36*(5), 492–504.

Belleflamme, P., Lambert, T., & Schwienbacher, A. (2010). Crowdfunding: An industrial organization perspective. In *Prepared for the Workshop Digital Business Models: Understanding Strategies' held in Paris on June 25–26*.

Belleflamme, P., Lambert, T., & Schwienbacher, A. (2014). Crowdfunding: Tapping the right crowd. *Journal of Business Venturing, 29*(5), 585–609.

Burtch, G., Ghose, A., & Wattal, S. (2015). The hidden cost of accommodating crowdfunder privacy preferences: A randomized field experiment. *Management Science, 61*(5), 949–962.

Butticè, V., Colombo, M. G., & Wright, M. (2017). Serial crowdfunding, social capital, and project success. *Entrepreneurship Theory and Practice, 41*(2), 183–207.

Cambridge Centre for Alternative Finance. (2017a, September). *Cultivating Growth—The 2nd Asia Pacific Region Alternative Finance Industry Report*.

Cambridge Centre for Alternative Finance. (2017b). *Hitting Stride—The Americas Alternative Finance Industry Report*.

Cambridge Centre for Alternative Finance. (2017c). *Entrenching Innovation—The 4th UK Alternative Finance Industry Report*.

Cambridge Centre for Alternative Finance. (2018). *Expanding Horizons—The 3rd European Alternative Finance Industry Report*.

Colombo, M. G., Franzoni, C., & Rossi-Lamastra, C. (2015). Internal social capital and the attraction of early contributions in crowdfunding. *Entrepreneurship Theory and Practice, 39*, 75–100.

Cumming, D., Leboeuf, G., & Schwienbacher, A. (2015). *Crowdfunding models: Keep-it-all vs. all-or-nothing*. Available at SSRN: https://ssrn.com/abstract=2447567 or http://dx.doi.org/10.2139/ssrn.2447567.

Cummins, M., Mac an Bhaird, C., Rosati, P., & Lynn, T. (2018). *Comparative evidence on the performance of non-bank financial institutions in business lending*. Available at SSRN: https://ssrn.com/abstract=3137177.

Davis, B. C., Hmieleski, K. M., Webb, J. W., & Coombs, J. E. (2017). Funders' positive affective reactions to entrepreneurs' crowdfunding pitches: The influence of perceived product creativity and entrepreneurial passion. *Journal of Business Venturing, 32*(1), 90–106.

Di Pietro, F., Masciarelli, F., & Prencipe, A. (2018a). *Believe or not believe? The effect of religiosity on individuals' participation on reward-based crowdfunding projects* (Working Paper). Luiss Business School.

Di Pietro, F., Spagnoletti, P., & Prencipe, A. (2018b). Fundraising across digital divide: Evidences from charity crowdfunding. In A. Lazazzara, R. C. D. Nacamulli, C. Rossignoli, & S. Za (Eds.), *Organizing in the digital economy. At the interface between social media, human behaviour and inclusion* (pp. 1–10). Cham, Switzerland: LNISO—Springer.

Di Pietro, F., Prencipe, A., & Majchrzak, A. (2018c). Crowd equity investors: An underutilized asset for open innovation in startups. *California Management Review, 60*(2), 43–70.

Di Pietro, F., Masciarelli, F., Prencipe, A., & Vangelis, S. (2018d). *The language of early stage investments: Venture capital and crowdfunding* (Working Paper).

Gerber, E. M., Hui, J. S., & Kuo, P. Y. (2012). Crowdfunding: Why people are motivated to post and fund projects on crowdfunding platforms. In *Proceedings of the International Workshop on Design, Influence, and Social Technologies: Techniques, Impacts and Ethics* (Vol. 2, No. 11).

Giudici, G., Massimiliano, G., & Rossi-Lamastra, C. (2018). Reward-based crowdfunding of entrepreneurial projects: The effect of local altruism and localized social capital on proponents' success. *Small Business Economics, 50*(2), 307–324.

Gleasure, R., & Feller, J. (2016). Emerging technologies and the democratisation of financial services: A metatriangulation of crowdfunding research. *Information and Organization, 26*(4), 101–115.

Gorbatai, A. D., & Nelson, L. (2015). Gender and the language of crowdfunding. In *Academy of Management Proceedings* (Vol. 2015, No. 1, p. 15785). Briarcliff Manor, NY: Academy of Management.

Haas, P., Blohm, I., & Leimeister, J. M. (2014). An empirical taxonomy of crowdfunding intermediaries. *International Conference on Information Systems (ICIS)*.

Horvát, E. Á., Uparna, J., & Uzzi, B. (2015). Network vs market relations: The effect of friends in crowdfunding. In *Proceedings of the 2015 IEEE/ACM International Conference on Advances in Social Networks Analysis and Mining 2015* (pp. 226–233).

Kgoroeadira, R., Burke, A., & van Stel, A. (2018). Small business online loan crowdfunding: Who gets funded and what determines the rate of interest? *Small Business Economics*, 1–21. Online first.

Lynn, T., Rosati, P., Nair, B., & Mac an Bhaird, C. (2017). Harness the crowd: An exploration of the #crowdfunding community on Twitter. *ISBE Annual Meeting 2017*. Belfast, UK.

Meer, J. (2014). Effects of the price of charitable giving: Evidence from an online crowdfunding platform. *Journal of Economic Behavior & Organization, 103*, 113–124.

Mollick, E. (2014). The dynamics of crowdfunding: An exploratory study. *Journal of Business Venturing, 29*(1), 1–16.

Morduch, J. (1999). The microfinance promise. *Journal of Economic Literature, 37*(4), 1569–1614.

Nevin, S., Gleasure, R., O'Reilly, P., Feller, J., Li, S., & Cristoforo, J. (2017). Social identity and social media activities in equity crowdfunding. In *Proceedings of the 13th International Symposium on Open Collaboration* (p. 11).

Parhankangas, A., & Renko, M. (2017). Linguistic style and crowdfunding success among social and commercial entrepreneurs. *Journal of Business Venturing, 32*(2), 215–236.

Paschen, J. (2017). Choose wisely: Crowdfunding through the stages of the startup life cycle. *Business Horizons, 60*(2), 179–188.

Reynolds, P. D. (2011). New firm creation: A global assessment of national, contextual and individual factors. *Foundations and Trends in Entrepreneurship, 6*(5–6), 315–496.

Schwienbacher, A., & Larralde, B. (2010). *Crowdfunding of small entrepreneurial ventures.* Available at: http://www.em-a.eu/fileadmin/content/REALISE_IT_2/REALISE_IT_3/CROWD_OUP_Final_Version.pdf. Last accessed 16 August 2018.

Signori, A., & Vismara, S. (2018). Does success bring success? The post-offering lives of equity-crowdfunded firms. *Journal of Corporate Finance, 50*, 575–591.

Smith, R. W., Faro, D., & Burson, K. A. (2013). More for the many: The influence of entitativity on charitable giving. *Journal of Consumer Research, 39*(5), 961–976.

Zhang, Y. (2012). An empirical study into the field of crowdfunding. *Economic Policy, 34*, 231–269.

CHAPTER 2

Addressing Information Asymmetries in Online Peer-to-Peer Lending

Mark Cummins, Theo Lynn, Ciarán Mac an Bhaird and Pierangelo Rosati

Abstract Digital technologies are transforming how small businesses access finance and from whom. This chapter explores online peer-to-peer (P2P) lending, a form of crowdfunding that connects borrowers and lenders. Information asymmetry is a key issue in online peer-to-peer lending marketplaces that can result in moral hazard or adverse selection, and ultimately impact the viability and success of individual platforms.

The authors are listed in alphabetical order.

M. Cummins · T. Lynn (✉) · P. Rosati
DCU Business School, Dublin City University, Dublin, Ireland
e-mail: theo.lynn@dcu.ie

M. Cummins
e-mail: mark.cummins@dcu.ie

P. Rosati
e-mail: pierangelo.rosati@dcu.ie

C. Mac an Bhaird
Dublin City University, Dublin, Ireland
e-mail: ciaran.macanbhaird@dcu.ie

T. Lynn et al. (eds.), *Disrupting Finance*, Palgrave Studies
in Digital Business & Enabling Technologies,
https://doi.org/10.1007/978-3-030-02330-0_2

Both online P2P lending platforms and lenders seek to minimise the impact of information asymmetries through a variety of mechanisms. This chapter discusses the structure of online P2P lending platforms and reviews how the disclosure of hard and soft information, and herding can reduce information asymmetries. The chapter concludes with a discussion of further avenues for research.

Keywords P2P lending · Online P2P lending · Information asymmetry

2.1   INTRODUCTION

It is widely agreed that small businesses play a critical role in economic growth, regardless of country size or development, by providing employment and income to a broad range of citizens, supporting a wider eco-system of firms, and fostering innovation (OECD 2017a). Their viability, sustainability, and growth depend on access to strategic resources, not least finance. The supply and sourcing of financing is a perennial strategic challenge for small businesses the world over, exacerbated by their innate characteristics and market inefficiencies. Due to under-collateralisation, limited or no credit history, and lack of sophisticated financial statements (and the expertise to produce such statements), a higher level of default risk is typically attached to small businesses and as a result access to credit is limited (Bhide 2003; OECD 2013). This situation persists despite a consistent decrease in the costs of financing in recent years, partly as result of the aforementioned characteristics of small business but also as a result of supply-side lending policies by traditional credit sources during the recent recession (Mills and McCarthy 2014; OECD 2017b).

Digital technologies are transforming the business models and dramatically increasing access to markets for small businesses. In the same way, it is also transforming how these businesses access finance and from whom. This chapter explores online peer-to-peer (P2P) lending, a form of crowdfunding (see Chapter 1) that bypasses conventional intermediaries, processes, and requirements to connect borrowers and lenders (Yum et al. 2012). Information asymmetry is a key issue in P2P lending that can result in moral hazard or adverse selection (Akerlof 1970) and ultimately impact the viability and success of individual P2P lending platforms. Both P2P lending platforms and lenders seek to minimise the impact of information asymmetries through a variety of mechanisms, most notably by supplementing hard information with soft information, and herding. The remainder of this chapter discusses the structure of online P2P lending platforms in greater detail. Extant literature on information asymmetries and online peer-to-peer lending platforms is then discussed. The chapter concludes with a discussion of further avenues for research.

2.2   Online Peer-to-Peer Lending Platforms

Online P2P lending platforms represent a convergence of P2P lending and collective financing, enabled by an Internet-based platform. Both P2P lending and collective financing are not new ideas in themselves. New venture financing, in particular, often mobilises existing peers based on family, friendship, or professional social relationships with the entrepreneur(s) (Berger and Udell 1998; Kotha and George 2012; Robb and Robinson 2014). This type of funding is often referred to as insider or informal funding in contrast to formal finance, whether intermediated from an equity or debt perspective (Berger and Udell 1998). Similarly, collective financing has a long history. Haas et al. (2014) cite the funding of the Statue of Liberty's pedestal in the nineteenth century as an early example of collective financing. In contrast, online P2P lending platforms harness the power of the Internet to enable an online marketplace for microcredit funding that acts as an intermediary to connect individuals or businesses wishing to obtain a loan (borrowers) with individuals and institutions wishing to fund loans (lenders). Lenders may not necessarily have existing social relationships with the entrepreneurs, management, or the business, and are more likely strangers with no preexisting relationship. They may be individuals or organisations established to provide credit on a formal basis. As such, online P2P lending platforms may, and increasingly do, allow traditional credit institutions such as banks to invest.

Online P2P lending marketplaces are two-sided networks where a P2P lending platform enables interactions between the demand (the borrower) and supply (the lender) sides of the network.[1] As well as recruiting potential borrowers and lenders (market-making), the P2P lending platform sets the rules or terms of engagement between borrowers and lenders in the platform. Typically, registered borrowers post

---

[1] For the remainder of this chapter, 'P2P lending' refers to online P2P lending.

their funding requirements on the platform and provide a relatively limited amount of information for due diligence purposes. The amount of information potential borrowers are required to submit varies among different platforms. For individuals, this might include detail on income, employment, other debt, purpose of loan, and a personal statement; for businesses, this typically includes financial accounts, some form of statement of historic trading, along with details of the lending proposition. In some marketplaces, additional information or verification can be requested (see later discussion on groups). The P2P lending platform then makes a decision to list the loan request or not. The P2P lending platform collects and scores prospective borrowers individually or as a pool, typically using a proprietary credit scoring mechanism. The potential loan requests are then offered to the prospective lenders through the platform. The prospective lenders then can decide to make offers (bids) to meet the full or partial loan requirements at a specific interest rate, if any.[2] Depending on the platform's functionality, such bids can be made by the prospective lenders manually (and independently), in groups, or using automated rule-based tools for portfolio management offered by the platform. By allowing lenders to invest in multiple small loans or small parts of a loan, the platform offers them the opportunity to diversify their loan portfolio and associated risk. The loan is funded when the minimum loan requirements are met, i.e. loan amount and interest rate. Finally, once the loan has been granted, the platform facilitates the loan processing and repayments and continues to collect and analyse the data relating to the loan and borrower for use in future credit scoring. P2P platforms mainly generate income from (1) origination fees from the borrower deducted at loan disbursement, (2) repayment fees charged to the lender when the borrower pays a monthly statement, and (3) additional charges such as late fees, loan part trading fees etc.

While the P2P lending platform undertakes a variety of functions including market-making, loan processing, and community-building activities, they do not, as a rule, participate in lending decisions (Meyer 2007; Wang et al. 2009).[3] As they do not make lending decisions

---

[2] In some social lending platforms, the interest rate can be zero e.g. kiva.org.

[3] A variant of online peer-to-peer lending platforms, balance sheet business lending and balance sheet property lending, has emerged in recent years and involves the platform entity providing loans directly to businesses. It accounts for a very small proportion of the sector. This has been excluded from this chapter as there is typically no market for the loans per se and the platform and lender are one and the same.

(or collect deposits as in traditional banks), they have much lower trans-action and intermediation costs than conventional credit institutions–key drivers of interest margins (Maudos and Guevara 2004). Operating costs are minimised through the use of online automated systems, operating outside the banking regulatory system, and not carrying the loans on their books thus avoiding liabilities for loans (Serrano-Cinca et al. 2015). These lower costs are transferred across to both the supply and demand sides of the P2P network. Accordingly, borrowers are attracted to P2P lending platforms by transparency, rapid decision-making, the promise of non-collateralised loans often at competitive interest rates (Sviokla 2009; Wang et al. 2009; IOSCO 2017) and lenders are attracted by lower transaction costs, risk diversification, access to market, and higher poten-tial returns (Morse 2015; IOSCO 2017).

Given the strong incentives for all stakeholders in the P2P lending value network, it is unsurprising that the P2P lending segment has expe-rienced rapid and substantial growth since the launch of what is con-sidered the first online P2P lending platform, Zopa.com, in 2005. P2P lending platforms may be categorised in a variety of ways including busi-ness model (profit/not-for-profit), number of borrowers per loan (one-to-one/one-to-many), borrower type (consumer/business) or loan use (e.g. real estate financing). P2P lending is still at a relatively early stage of development and analyst reports on market sizing is characterised by regional focus, definitional ambiguity, and significant variances. Table 2.1 provides a summary of the size of the overall market and main segments by region.

The rapid growth of P2P lending has been justified through two main arguments–financial intermediation theory and market equilibrium the-ory (Serrano-Cinca et al. 2015). The financial intermediation hypothe-sis suggests that as P2P lending platforms are more cost efficient than traditional credit institutions and therefore have lower intermediation costs, they are more attractive to both lenders and borrowers for the reasons discussed earlier. The market equilibrium hypothesis recognises that if markets function efficiently, supply and demand should be in equi-librium. However, a credit rationing problem exists, particularly in eco-nomic downturns, in that some prospective borrowers may not receive loans even if they are willing to pay higher interest rates. Proponents of the market equilibrium hypothesis argue that P2P lending platforms solve this credit rationing problem and bring the credit market towards equilibrium.

Table 2.1   Size of the P2P lending market by region

Region	Total market size	P2P consumer lending	P2P business lending
		2015	
The Americas	28.70	18	2.6
Asia Pacific and China	108.85	52.78	39.99
Europe	1.108	0.398	0.23
Middle East and Africa	0.242	0.010	0.023
		2016	
The Americas	35.2	21.1	1.3
Asia Pacific and China	244.43	137.02	58.51
Europe	2.171	0.733	0.368
Middle East and Africa	0.36	0.033	0.031

Notes All figures are reported in USD/billions
Sources Cambridge Centre for Alternative Finance (2017a, b, 2018a, b)

2.3   INFORMATION ASYMMETRIES AND PEER TO PEER LENDING PLATFORMS

As discussed previously, except in a limited number of instances, for example, balance sheet business lending and balance sheet property lending, loans are granted by lenders and not the platform per se; the platform operator transfers the credit risk to the lenders. Addressing information asymmetries is a major theme of P2P lending platform research. As in the overwhelming majority of investment decisions, P2P lenders are at a disadvantage to the borrower with regard to the loan decision. The borrower has near-complete information while the lender has only what is presented by the P2P lending platform. As such, platform operators must design mechanisms into their platform and process to reduce these asymmetries while not demotivating either potential borrowers or lenders with unnecessary barriers to participate. Such mechanisms include provision point mechanisms (all or nothing), general platform rules, feedback systems, crowd due diligence, and safeguard funds. The provision of data and associated analysis and, in particular big data as online P2P lending marketplaces reach sufficient scale, is a critical component of differentiating P2P lending platforms but also reducing information asymmetries (Yan et al. 2015). A key mechanism in all P2P lending platforms is some form of categorisation of a loan, typically proprietary, based on some platform assessment of the creditworthiness of the borrower represented by a credit grade (if not a credit score) representing the likelihood that the borrower will repay their debt. Such scores are derived from mandatory information disclosures from prospective borrowers such as credit history and personal data, but also supplemental voluntary information disclosures including more detailed biographical data, photographs, and in some cases, peer endorsements (Yan et al. 2015). Both conventional and non-standard information has been explored by researchers in terms of their contribution to key outcomes of the lending process, e.g. loan funding, final interest rate level, and default.

The importance of the credit grade as a signal is underlined by extant research which suggests, unsurprisingly, that higher credit grades are predictive of successful loan funding and lower risk of default (Greiner and Wang 2010; Emekter et al. 2015). A significant focus of research has been on so-called hard information which is easy to compare across borrowers and categories of borrowers. In addition to credit grade, debt-income ratio, bank account verification, and borrower debt level have been found to be significant for predicting funding probability and final interest rate (Greiner and Wang 2010; Serrano-Cinca et al. 2015). Despite this, it should be noted that Freedman and Jin (2008) suggest that exposure to credit grades rather than actual credit scores adversely effects loan decision making. In a similar vein, recent research suggests that credit grades may not represent accurate estimates of borrowers' creditworthiness, and that the accuracy of hard information for P2P lending decision-making is improved with further information disclosure (Serrano-Cinca et al. 2015; Tao et al. 2017; Zhu 2018).

Soft information in the context of P2P lending research typically refers to information about the borrower and their individual situation (Dorfleitner et al. 2016). Soft information is often viewed as a means to addressing information asymmetry and associated adverse selection in P2P lending platforms (Weiss et al. 2010; Gao et al. 2016; Prystav 2016). Iyer et al. (2009) suggest that screening through soft information, in this case, the number of friend endorsements and the loan purpose, is relatively more important when evaluating lower-quality borrowers. This is consistent with Prystav (2016) who found that borrowers with poorer relative credit ratings will be ignored if not for soft information. Gao et al. (2016) also identify that loan purpose is taken into account by lenders but they also note that they may be deceived by such information. Several studies have examined the narrative descriptions provided in loan listings. Research by Pötzsch and Böhme (2010) suggests that communicating soft information relating to the borrower's education, profession, and qualifications had a small but significant effect. Research by Herzenstein et al. (2011b) suggest that borrower claims about themselves (identities) in narratives that focus on trustworthiness or success increase the likelihood of loan funding but have less predictive loan performance than other identities, e.g. borrowers claiming economic hardship. Furthermore, borrowers who claim more identities in narratives both have increased likelihood of loan funding and a reduced final interest rate. Similarly, Michels (2012) suggests that voluntary information disclosures, over and above those required by the platform, even when unverified, results in an increase in bidding activity by prospective lenders and a reduction of interest rates. Dorfleitner et al. (2016) compared two European P2P lending platforms, and found that description text, and specifically spelling errors, text length, and the sentiment intensity of keywords, predicted funding probability on the less restrictive of the two platforms examined. Gao et al. (2016) show that the presence of well-established features that influence reader behaviours (readability, positive tones, and deception cues) in narrative texts of loan listings all meaningfully relate to loan repayment.

Many researchers have explored the extent to which lenders are rational or perceptual in their decision-making on P2P lending platforms. Research on the Prosper.com platform by Herzenstein et al. (2008) suggests that prospective borrowers who provided a photo affected loan success negatively but found while demographic attributes, such as race and gender do affect likelihood of funding success, their influence was minor compared to other factors. Pope and Sydnor (2011) examining the same platform found evidence of racial disparities with loan listings featuring pictures with 'blacks' 25–35% less likely to receive funding than pictures featuring 'whites'. Similarly, Ravina (2012) reports that after hard information is taken into account, more attractive prospective borrowers have a higher likelihood of loan funding and lower interest rates and that consistent with Pope and Sydnor (2011) identifies disparities between 'blacks' and 'whites'. Again using photographs of borrowers from Prosper.com, Duarte et al. (2012) found that borrowers who appear more trustworthy have higher probabilities of having their loans funded, have better credit scores and default less often. Age-based research would seem to be conclusive. Gonzalez and Loureiro (2014) found that loan success is sensitive to relative age and attractiveness. Furthermore, they found that (a) attractiveness had no effect where perceived age might signal competence, and (b) when the lender and borrower were of the same gender, attractiveness might negatively impact loan success. Similarly, research on Chinese borrower perceptions by Chen et al. (2016) suggests that borrowers perceived having a shared birthplace, location or occupation with lenders increased the 'ease of funding'. Burtch et al. (2014) note that location proximity and cultural differences in borrower and lender country of origin impact loan funding.

In many, but not all, P2P lending platforms operate auctions where the loan is only funded on a 'fund it all' basis, i.e. there must be sufficient bids to fund the loan amount requested loan in its entirety. This is sometimes referred to as the "rule of full funding" (Herzenstein et al. 2011a). As per other financial markets, where there is asymmetric (or imperfect) information, investors tend to herd (Bikhchandani and Sharma 2000). Bikhchandani and Sharma (2000, p. 280) define herding as follows:

> an individual can be said to herd if she would have made an investment without knowing other investors' decisions, but does not make that investment when she finds that others have decided not to do so. Alternatively, she herds when knowledge that others are investing changes her decision from not investing to making the investment.

Bikhchandani and Sharma (2000) differentiate between two types of herding–intentional and spurious. The former occurs when one set of investors copies another set of investors behaviour intentionally, i.e. the mimicry is post hoc; the latter occurs when investors behave similarly whether they are aware of the others investors' behaviour or not. Intentional herding may be rational or irrational. Rational herding is based on the observation of publicly visible investment choices or actions by one or more investors and therefore involves some form of observational learning and information cascades. In contrast, irrational whereas irrational herding is based on irrational beliefs or sentiment and is typified by momentum-investment strategies.

In the context of P2P lending, lenders may be particularly prone to herding due to the transparent nature of the platforms. Indeed this transparency and access to data also makes such platforms a fertile space for academic research. Berkovich (2011) analyses data from Prosper. com and his findings suggest that there is evidence of herding as per the model in Berkovich and Tayon (2009). Herzenstein et al. (2011a) examine strategic herding behaviour by lenders, again on Prosper.com. They identify that lenders are likely to herd in active auctions up until the loan is fully funded at which point, herding behaviour decreases. Interestingly, Herzenstein et al. (2011a) find that there is a positive association between strategic herding and loan repayment and suggest that such behaviour therefore benefits lenders individually and collectively. Zhang and Liu (2012) explore Prosper.com data also and conclude that the lenders engage in both rational and irrational herding based on the evidence of observational learning. Zhang and Liu's study is noteworthy as they observed counterintuitive herding effects, e.g. low credit scores amplified herding effects whereas favourable borrower characteristics seem to dampen herding effects. They also found that rational herding outperformed irrational herding in predicting loan performance.

Lee and Lee (2012) explore herding behaviour on a Korean P2P lending platform, Popfunding.com. Again, they find strong evidence of herding behaviour including a diminishing marginal effect of the observed herding behaviour as per Herzenstein et al. (2011a). More recently, Zhang and Chen (2017) investigate herding on a Chinese P2P lending platform. Again they find evidence of herding and in this case are able to identify both rational and irrational herding behaviour.

As a final related point, it is worth noting research relating to borrower groups within P2P lending platforms. In some P2P lending platforms, such as Prosper.com, it is common for borrowers to form groups comprising other borrowers, who may also in themselves act as lenders. Group leaders may set membership criteria that can require additional information from members over and above that required by the P2P lending platform and which may only be available to group leaders or the group. This group-specific private information is not available to all participants in the P2P lending platform. Given their status in groups, group leaders may wield considerable influence through endorsements or leading bidding. Research is inconclusive on groups (Lee and Lee 2012). Everett (2015) suggests that membership in a group with private information or enhanced monitoring is associated with lower default rates however not necessarily lower interest rates (Everett 2015). Everett (2015) suggests this private information disclosure, while solving an information asymmetry problem for some lenders, introduces a hold-up problem for some borrowers. Everett (2015) finds that consistent with extant literature, the interest rate often depends on the social relationship with the prospective borrower and the quality of the credit rating with more professional lenders seeking higher economic rents through higher interest rates. Notwithstanding this, Chen et al. (2016) find that group membership and the borrower's credibility and trust within that group yielded inconsistent results, however, the degree of group inclusiveness had a negative impact on, his/her funding and repayment performance.

Borrower groups may also play a role in herding. For example, group leader endorsements and bidding can initiate cascades leading to herding. Regarding the impact of group leader actions on loan funding, Kumar (2007) suggests group leader endorsement can increase the likelihood of loan funding success. However, the impact of group leader behaviour on interest rates is less conclusive. For example, while Berger and Gleisner (2010) suggest that active bidding by the group leader with others, and in itself, may result in lower interest rates, Freedman and Jin (2008) suggest that in certain instances group leader actions will increase the average interest rate.

2.4 Conclusions and Future Directions for Research

In this chapter, we provided an overview of online P2P lending platforms and discussed the extant literature on how information asymmetries are reduced through various platform mechanisms and the lenders themselves, including information disclosure, herding and relatedly in-platform groups. Peer-to-peer lending as a subset of the wider crowdfunding and FinTech domain is experiencing rapid adoption worldwide and is the dominant segment of most alternative finance markets. While there is a substantial literature base on information asymmetries and P2P lending, the increasingly global adoption of P2P lending, the proliferation of new platforms and marketplaces, and the evolution of new technologies provides a fertile ground for future research which we will discuss briefly.

Researchers have suggested that understanding the behaviour, and in particularly the 'inner life', of investors requires a greater appreciation of the both the socio-economic and technical context in which investment takes place (Hirsto 2011; Zwick and Schroeder 2011). For example, US data prior to 2008 operated under a different regulatory environment when the SEC required registration under the Securities Act of 1933 resulting in changes to platform operation. Similarly, different countries operate under different levels of regulation and oversight. While there has been a small number of European studies and an increasing base of literature from China, the majority of early research has a US-focus. There is little truly comparative work examining the impact of local socio-economic forces, culture, language, and other aspects of national identity on borrower and lender behaviour on domestic and international P2P lending platforms.

The number of P2P lending platforms has increased dramatically since 2005. In China alone, media reports suggest over 2000 P2P lending platforms were active in the market in February 2018. Today, there are P2P lending platforms of every hue; they each have different features, functionalities, and affordances that impact the operation of the market. As with geo-cultural focus of early research, early P2P lending research focused on available datasets such as Prosper.com. Whether extant findings are generalisable across all platforms and take into account platform idiosyncrasies is worthy of further exploration.

The role and impact of information disclosure, hard and soft, has been the focus of much of the academic research to date. In this era of big data and API-fication, platforms are increasingly looking to integrate third-party data sources into P2P lending platforms. Yan et al. (2015) discuss the potentially transformational role such big data can have in reducing information asymmetry through reduced signalling and search costs. While highlighting the benefits of increased data volumes and variety, they warn that such volumes and velocity of data will only reduce information asymmetries where the quality of the data analysis and subsequent analysis is relatively high. Related technologies such as machine learning, deep learning and artificial intelligence similarly can contribute to loan decision-making and reducing information asymmetries but provide their own unique challenges and may result in unexpected consequences, not least diluting or removing the human element in P2P lending.

This chapter primarily focuses on information disclosure and herding as a means of reducing information asymmetries in P2P lending. Due to the large number of platforms operating today, there is greater heterogeneity in the structural mechanisms for reducing information asymmetries including provision point mechanisms, platform rules, contractual agreements, etc. The heterogeneity, scale, and global footprint of P2P lending and the strategic necessity of reducing information asymmetries to ensure the efficient operation of P2P lending platforms ensure a fervent space for scholarly inquiry and impact.

REFERENCES

Akerlof, G. A. (1970). The market for lemons: Quality uncertainty and the market mechanism. *Quarterly Journal of Economics, 84*(3), 488–500.

Berger, S. C., & Gleisner, F. (2010). Emergence of financial intermediaries in electronic markets: The case of online P2P lending. *BuR Business Research Journal, 2*(1). Available at SSRN: https://ssrn.com/abstract=1568679.

Berger, A. N., & Udell, G. F. (1998). The economics of small business finance: The roles of private equity and debt markets in the financial growth cycle. *Journal of Banking & Finance, 22*(6), 613–673.

Berkovich, E. (2011). Search and herding effects in peer-to-peer lending: Evidence from prosper.com. *Annals of Finance, 7*(3), 389–405.

Berkovich, E., & Tayon, R. (2009). Herding and crowding to efficiency. Phd. Dissertation—Essays on Search and Herding. University of Pennsylvania, Pennsylvania.

Bhide. (2003). *The origin and evolution of new business.* New York: Oxford University Press.

Bikhchandani, S., & Sharma, S. (2000). Herd behavior in financial markets. *IMF Staff Papers, 47*(3), 279–310.

Burtch, G., Ghose, A., & Wattal, S. (2014). Cultural differences and geography as determinants of online prosocial lending. *Management Information Systems Quarterly, 38*(3), 773–794.

Cambridge Centre for Alternative Finance. (2017a). *Hitting Stride—The Americas Alternative Finance Industry Report.*

Cambridge Centre for Alternative Finance. (2017b, September). *Cultivating Growth—The 2nd Asia Pacific Region Alternative Finance Industry Report.*

Cambridge Centre for Alternative Finance. (2018a). *Expanding Horizons—The 3rd European Alternative Finance Industry Report.*

Cambridge Centre for Alternative Finance. (2018b, June). *The 2nd Annual Middle East and Africa Alternative Finance Industry Report.*

Chen, X., Zhou, L., & Wan, D. (2016). Group social capital and lending outcomes in the financial credit market: An empirical study of online peer-to-peer lending. *Electronic Commerce Research and Applications, 15*, 1–13.

Dorfleitner, G., Priberny, C., Schuster, S., Stoiber, J., Weber, M., de Castro, I., & Kammler, J. (2016). Description text related soft information in peer-to-peer lending—Evidence from two leading European platforms. *Journal of Banking & Finance, 64*, 169–187.

Duarte, J., Siegel, S., & Young, L. (2012). Trust and credit: The role of appearance in peer-to-peer lending. *The Review of Financial Studies, 25*(8), 2455–2484.

Emekter, R., Tu, Y., Jirasakuldech, B., & Lu, M. (2015). Evaluating credit risk and loan performance in online peer-to-peer (P2P) lending. *Applied Economics, 47*(1), 54–70.

Everett, C. R. (2015). Group membership, relationship banking and loan default risk: The case of online social lending. *Banking and Finance Review, 7*(2), 15–54.

Freedman, S., & Jin, G. Z. (2008). *Do social networks solve information problems for peer-to-peer lending? Evidence from Prosper.com* (NET Institute Working Paper No. 08-43). Available at SSRN: https://ssrn.com/abstract=1304138.

Gao, Q., Lin, M., & Sias, R. W. (2016). *Words matter: The role of texts in online credit markets.* Available at SSRN: https://ssrn.com/abstract=2446114.

Gonzalez, L., & Loureiro, Y. K. (2014). When can a photo increase credit? The impact of lender and borrower profiles on online peer-to-peer loans. *Journal of Behavioral and Experimental Finance, 2,* 44–58.

Greiner, M. E., & Wang, H. (2010). Building consumer-to-consumer trust in e-finance marketplaces: An empirical analysis. *International Journal of Electronic Commerce, 15*(2), 105–136.

Haas, P., Blohm, I., & Leimeister, J. M. (2014). An empirical taxonomy of crowdfunding intermediaries. *Proceedings of Thirty Fifth International Conference on Information Systems.* Auckland: AIS.

Herzenstein, M., Dholakia, U. M., & Andrews, R. L. (2011a). Strategic herding behavior in peer-to-peer loan auctions. *Journal of Interactive Marketing, 25*(1), 27–36.

Herzenstein, M., Sonenshein, S., & Dholakia, U. M. (2011b). Tell me a good story and I may lend you money: The role of narratives in peer-to-peer lending decisions. *Journal of Marketing Research, 48*(SPL), S138–S149.

Herzenstein, M., Andrews, R. L., Dholakia, U. M., & Lyandres, E. (2008). *The democratization of personal consumer loans? Determinants of success in online peer-to-peer lending communities.* Boston University School of Management Research Paper, 14(6).

Hirsto, H. (2011). Everyday discourses of stock market investing: Searching for investor power and responsibility. *Consumption, Markets and Culture, 14*(1), 57–77.

IOSCO. (2017, February). *IOSCO Research Report on Financial Technologies (Fintech).*

Iyer, R., Khwaja, A. I., Luttmer, E. F., & Shue, K. (2009). *Screening in new credit markets: Can individual lenders infer borrower creditworthiness in peer-to-peer lending?* (AFA 2011 Denver Meetings Paper). Available at SSRN: https://ssrn.com/abstract=1570115.

Kotha, R., & George, G. (2012). Friends, family, or fools: Entrepreneur experience and its implications for equity distribution and resource mobilization. *Journal of Business Venturing, 27*(5), 525–543.

Kumar, S. (2007). Bank of one: Empirical analysis of peer-to-peer financial marketplaces. *AMCIS 2007 Proceedings* (p. 305).

Lee, E., & Lee, B. (2012). Herding behavior in online P2P lending: An empirical investigation. *Electronic Commerce Research and Applications, 11*(5), 495–503.

Maudos, J., & De Guevara, J. F. (2004). Factors explaining the interest margin in the banking sectors of the European Union. *Journal of Banking & Finance, 28*(9), 2259–2281.

Meyer, T. (2007, July). Online P2P lending nibbles at banks' loan business. *Deutsche Bank Research.*

Michels, J. (2012). Do unverifiable disclosures matter? Evidence from peer-to-peer lending. *The Accounting Review, 87*(4), 1385–1413.

Mills, K., & McCarthy, B. (2014). *The state of small business lending* (Technical Report, Harvard Business School Working Paper).

Morse, A. (2015). Peer-to-peer crowdfunding: Information and the potential for disruption in consumer lending. *Annual Review of Financial Economics, 7*, 463–482.

OECD. (2013). *SME and entrepreneurship financing: The role of credit guarantee schemes and mutual guarantee societies in supporting finance for small and medium-sized enterprises.* Paris: OECD Publishing. Available at: https://one.oecd.org/document/CFE/SME(2012)1/FINAL/en/pdf. Last accessed 18 July 2018.

OECD. (2017a). *Enhancing the contributions of SMEs in a global and digitalised economy.* Paris: OECD Publishing. Available at: https://www.oecd.org/mcm/documents/C-MIN-2017–8-EN.pdf. Last accessed 18 July 2018.

OECD. (2017b). *Financing SMEs and entrepreneurs 2017. An OECD scoreboard.* Paris: OECD Publishing. Available at: https://www.oecd.org/cfe/smes/Financing%20SMEs%20and%20Entrepreneurs%202017_Highlights.pdf. Last accessed 18 July 2018.

Pope, D. G., & Sydnor, J. R. (2011). What's in a picture? Evidence of discrimination from Prosper. com. *Journal of Human Resources, 46*(1), 53–92.

Pötzsch, S., & Böhme, R. (2010). The role of soft information in trust building: Evidence from online social lending. In *International Conference on Trust and Trustworthy Computing* (pp. 381–395). Berlin and Heidelberg: Springer.

Prystav, F. (2016). Personal information in peer-to-peer loan applications: Is less more? *Journal of Behavioral and Experimental Finance, 9,* 6–19.

Ravina, E. (2012). Love & loans: The effect of beauty and personal characteristics in credit markets (June 15, 2018). Available at SSRN: https://ssrn.com/abstract=1107307 or https://doi.org/10.2139/ssrn.1107307.

Robb, A. M., & Robinson, D. T. (2014). The capital structure decisions of new firms. *Review of Financial Studies, 27*(1), 153–179.

Serrano-Cinca, C., Gutierrez-Nieto, B., & López-Palacios, L. (2015). Determinants of default in P2P lending. *PLoS One, 10*(10), e0139427.

Sviokla, J. (2009). Breakthrough ideas: Forget Citibank—Borrow from Bob. *Harvard Business Review, 87*(2), 29–40.

Tao, Q., Dong, Y., & Lin, Z. (2017). Who can get money? Evidence from the Chinese peer-to-peer lending platform. *Information Systems Frontiers, 19*(3), 425–441.

Wang, H., Greiner, M., & Aronson, J. E. (2009). People-to-people lending: The emerging e-commerce transformation of a financial market. *Value creation in E-business management* (pp. 182–195). Berlin and Heidelberg: Springer.

Weiss, G. N. F., Pelger, K., & Horsch, A. (2010). Mitigating adverse selection in P2P Lending—Empirical evidence from prosper.com (July 29, 2010). Available at SSRN: https://ssrn.com/abstract=1650774 or https://doi.org/10.2139/ssrn.1650774.

Yan, J., Yu, W., & Zhao, J. L. (2015). How signaling and search costs affect information asymmetry in P2P lending: The economics of big data. *Financial Innovation, 1*(1), 19.

Yum, H., Lee, B., & Chae, M. (2012). From the wisdom of crowds to my own judgment in microfinance through online peer-to-peer lending platforms. *Electronic Commerce Research and Applications, 11*(5), 469–483.

Zhang, K., & Chen, X. (2017). Herding in a P2P lending market: Rational inference OR irrational trust? *Electronic Commerce Research and Applications, 23*, 45–53.

Zhang, J., & Liu, P. (2012). Rational herding in microloan markets. *Management Science, 58*(5), 892–912.

Zhu, Z. (2018). Safety promise, moral hazard and financial supervision: Evidence from peer-to-peer lending. *Finance Research Letters*.

Zwick, D., & Schroeder, J. (2011). Stock trading in the digital age: Speed, agency, and the entrepreneurial consume. In R. W. Belk & R. Llamas (Eds.), *The Routledge companion to digital consumption*. Abingdon, UK: Routledge.

CHAPTER 3

Machine Learning and AI for Risk Management

Saqib Aziz and Michael Dowling

Abstract We explore how machine learning and artificial intelligence (AI) solutions are transforming risk management. A non-technical overview is first given of the main machine learning and AI techniques of benefit to risk management. Then a review is provided, using current practice and empirical evidence, of the application of these techniques to the risk management fields of credit risk, market risk, operational risk, and compliance ('RegTech'). We conclude with some thoughts on current limitations and views on how the field is likely to develop in the short- to medium-term. Overall, we present an optimistic picture of the role of machine learning and AI in risk management, but note some practical limitations around suitable data management policies, transparency, and lack of necessary skillsets within firms.

S. Aziz · M. Dowling (✉)
Rennes School of Business, Rennes, France
e-mail: michael.dowling@rennes-sb.com

S. Aziz
e-mail: saqib.aziz@rennes-sb.com

© The Author(s) 2019
T. Lynn et al. (eds.), *Disrupting Finance*, Palgrave Studies
in Digital Business & Enabling Technologies,
https://doi.org/10.1007/978-3-030-02330-0_3

Keywords AI · Machine learning · Risk management · RegTech · Credit risk · Operational risk · Market risk

3.1   Introduction

Artificial intelligence (AI), and the machine learning techniques that form the core of AI, are transforming, and will revolutionise, how we approach financial risk management. Everything to do with understanding and controlling risk is up for grabs through the growth of AI-driven solutions: from deciding how much a bank should lend to a customer, to providing warning signals to financial market traders about position risk, to detecting customer and insider fraud, and improving compliance and reducing model risk. In this chapter we detail current machine learning and AI techniques being used and current applications of those techniques. We further envisage the future role for fully AI solutions as the natural next step after the widespread adoption of machine learning in helping the organisation to manage risk.

An example of ZestFinance serves to illustrate the potential for AI and machine learning in risk management. ZestFinance was founded by a former Chief Information Officer of Google and in 2016 partnered with Baidu, the dominant search engine in China, to improve Baidu's lending decisions in the Chinese market. Baidu was particularly interested in making small loan offers to retail customers buying products from their platform. Unlike most developed countries, the risk with lending in the Chinese market is that less than 20% of people have credit profiles or credit ratings. Lending to people who have either 'thin' credit profiles, or no credit profiles, is inherently risky as there is no history to draw on to check borrower reliability. ZestFinance (with permission) taps into the huge volume of information on members held by Baidu such as their search or purchase histories to help Baidu decide whether to lend. They use thousands of data points per customer and are still able to make lending decisions on new applications in seconds. A reported trial in 2017 of their system led to a 150% increase in total small-item lending by Baidu with no increase in credit losses in the space of just two months.[1]

The exact nature of how ZestFinance makes these decisions is not disclosed except under the broad umbrella of machine learning and AI, but essentially what they use as a base is a core machine learning set of

---

[1] https://www.technologyreview.com/s/603604/an-ai-fueled-credit-formula-might-help-you-get-a-loan/.

techniques around clustering and decision trees, and possibly deep learning. Thus a search history indicating accessing gambling websites would cluster a potential borrower into a higher risk group. While on the other hand, a history of online spending that supports the applicant's reported income in their application to borrow, or maybe even data indicating upward career mobility, might cluster someone into a group of lower risk borrowers. The actual clusters will be significantly more refined than the simple examples given above. This approach crosses the line from machine learning to AI due to the automated nature of the lending decision-making process.

The ZestFinance example illustrates the essence of how AI and machine learning can improve risk management. A standard credit score is largely a linear calculation of a small number (about 50) positive or negative numerical characteristics about a person and thus misses out on a huge amount of additional personal information that can help to either reduce negative risk or accept positive risk. This new information is often atypical and non-numerical; the type of data which machine learning is innately suited to analysing. Our chapter thus expands on this idea of how AI and machine learning can improve risk management—particularly around the techniques being used to make decisions based on such large volumes of atypical data. The next section briefly details, in a non-technical manner, the core machine learning techniques which can be applied to improve risk management. The remainder of the chapter is devoted to the actual application of AI and machine learning to various forms of risk management, finishing with a forward-looking perspective on what is next for the role of AI in risk management and some challenges that need to be addressed.

3.2   Machine Learning and AI Techniques for Risk Management

A first step is defining what we mean by AI and machine learning, and this is not necessarily a straightforward distinction. In a glib sense the public relations and fundraising functions of startups tend to use the more attractive AI term when they most often mean machine learning, but even in research there is a reasonably fluid distinction. AI is most commonly viewed as intelligence demonstrated by machines, with intelligence being defined with reference to what we view intelligence as in humans (Turing 1952, cf. Shieber 2004). As it matters to risk management we are normally particularly interested in artificial super-intelligence; that is machines that can demonstrate a risk management-specific intelligence higher than human intelligence in this field. To compare the two terms in a more technical manner, we can say machine learning is a core technique of AI involving learning from data, but that AI often involves additional techniques and requirements. For example, as noted by Bostrom (2014), a full AI solution would be automated in terms of data identification, data testing, and making decisions based on the data testing. In practice, AI might involve additional techniques in addition to machine learning, such as including hard-coded and logic rules. Machine learning on the other hand normally involves manual data identification and testing by the data scientist, and human decisions as to how to apply the outputted information. Given the lack of technological and organisational readiness for pure AI, and the reality that most claimed AI is in fact machine learning, in this section we outline the core machine learning techniques applied to risk management. In the following section and especially the last section, we move our discussion more towards AI as the logical next step to follow from the widespread usage of machine learning techniques.

Machine learning falls into two broad categories of supervised and unsupervised machine learning. In supervised learning you have input data that you wish to test to determine an output. This is similar to how in traditional statistics terms you have a range of independent variables that you test to determine relationship with the dependent variable. In unsupervised learning, you only have input data and wish to learn more about the structure of the data. Table 3.1 shows the distinction between these two categories as well as the main techniques within each category. A category of note that crosses both supervised and unsupervised learning is deep learning, which is viewed as the closest towards AI, and to which we turn at the end of this section.

Regression machine learning is the closest group of techniques to that usually applied in traditional determination of the causal relationship between variables. In simple terms, we might describe a traditional linear regression equation for a credit lending risk assessment as perhaps the dependent variable being the risk of loan non-repayment, which is then sought to be explained by a range of independent variables that should influence the risk of loan non-repayment. These independent variables might, for example, include financial measures such average non-repayment rates, whether the person is full-time employed, whether they have a good credit history, and whether they own property.

3 MACHINE LEARNING AND AI FOR RISK MANAGEMENT   37

Table 3.1 Categories of machine learning techniques (*Source* van Liebergen 2017)

		Linear methods	Non-linear methods
Problem type	Supervised Regression	• Principal components • Ridge • Partial least squares • LASSO	Penalised regression: • LASSO • LARS • Elastic nets Neural networks and deep leaning
	Classification	Support vector machines (SVM)	Decision trees: • Classification trees • Regression trees • Random forest SVM Deep learning
	Unsupervised Clustering[a]	Clustering methods: K- and X-means, hierarchical principal components analysis Deep learning	

[a]Since unsupervised methods do not describe a relation between a dependent and interdependent variable, they cannot be labelled linear or non-linear

Regression machine learning differs from traditional regression in that it uses regression techniques that allow for large numbers of variables to be used as independent variables and then automatically discarded if they lack explanatory power. This is a necessary feature due to the large range of data that is available to the data scientist. It also reduces the extent of theorising needed to determine suitable independent variables. Thus, LASSO regression zero weights independent variables with low explanatory power, while Ridge regression gives lower weights to variables in a model that are highly correlated with other variables in a model. In both cases the outcome is a reduced model that allows the data scientist to move from large numbers of potential explanatory variables to a smaller subset. A LARS regression works in the opposite direction to a LASSO and Ridge, by initially zero-weighting all variables and only adding variables that are shown to have explanatory power. Hu et al. (2015) provide an interesting application of this set of techniques by applying a LASSO regression technique, among others, to the development of an automated investment trading advice system based on stock trend analysis and showing this method has improved trading performance compared to traditional methods.

Principal Component Analysis (PCA) and partial-least squares regressions are quite similar in that they both aim to reduce the number of variables by combining variables and extracting common factors. PCA is the more popular of the two as it is widely used in traditional statistics and therefore better understood. A simple example of PCA is that for a set of potential variables to be used to determine credit repayment risk consisting of: (1) owns a house, (2) owns a car, and (3) has significant savings: a common factor might be extracted from these that could be termed 'asset ownership'. A more applied example is provided by Zhong and Enke (2017) who use PCA to reduce 60 correlated economic and financial measures to a smaller set of factors and then apply a set of neural network machine learning techniques to forecast the US S&P 500 index. This latter example is a common use of PCA in machine learning: that it is used as a first step for dimension reduction and then a more advanced machine learning technique is applied to learn from the PCA components.

The other main category of supervised learning is 'classification', with support vector machines (SVM) and decision trees being the most popular techniques within this group. The outcomes and visuals of decision trees are easily understood in practice and therefore amenable to explanation to non-data scientists. A famous application of the decision tree technique is to Titanic survival prospects. Starting with the initial tree trunk statistic that 62% of passengers died and 38% survived the Titanic disaster, we can build a decision tree to classify groups that had a greater or lesser chance of survival. Thus, a first classification of whether a passenger is male or female shows that only 19% of males survived while 74% of females survived. We can branch out the tree even further by, for example, looking at age groups beneath the first male–female division, and this shows that despite only 19% of males surviving, male boys under the age of 6 had a 67% survival chance.

This classification approach of creating sub-groups then helps to understand what characteristics determine outcomes. In practice decision trees for machine learning have many more challenges than in the Titanic example, particularly around the issue of overfitting to existing data and thus determined sub-groups having poor predictive power with new data and in new situations. SVM are more complex in formulation than decision trees but have the same essential end goal of creating groups based on input characteristics to classify and predict outcomes. In the case of SVM the approach is to map characteristics on a plane and classify groups based on similarity of where they are on this plane. As a financial risk example, Nazemi et al. (2018) apply a SVM approach to predict financial loss to holders of bonds that have defaulted (normally some proportion of loss is recoverable from firm assets after bond default) and are able to demonstrate the superiority of the approach compared to more conventional methods.

Turning now to unsupervised learning, we first have clustering machine learning techniques which have some similarity with SVM in that they involve mapping characteristics on a plane. The technique differs in that it is not trying to predict outcomes but instead create similar groups. Email 'spam' detection, for example, is usually based on a clustering approach—if an email looks like other emails that are deemed spam then it is likely to also be spam. K-means clustering is the most popular approach, although other methods such as X-means and hierarchical clustering are growing in popularity. We'll focus on K-means clustering to describe the general technique. In K-means a number of cluster groups the data scientist wishes to arrive at is predetermined (although in practice a range of the number of groups is tested), characteristics are mapped on a plane and a dividing line (not necessarily straight) is drawn that best distinguishes between groups. The idea of the iterative process behind the technique is to maximise the difference in means between determined groups.

Deep learning and neural networks[2] are viewed as being at the forefront of machine learning techniques and are often classified separately to the machine learning techniques already described. Indeed deep learning can be both supervised and unsupervised forms of learning depending on the purpose to which it is being applied. The intuition behind deep learning is to more accurately model complex relationships between variables and ultimately to better mimic human decision-making. To that extent these techniques represent the closest to actual AI techniques, albeit still missing some of the data identification and automation features necessary for true AI. Heaton et al. (2017) provide a detailed analysis of the recent application of deep learning to finance, as well as highlighting the potential of the approach.

A key feature of deep learning is the addition of 'hidden layers' after the input data stage that allow multiple and combined influences

---

[2]While subject to some debate, it is worth thinking of deep learning as a newer term of neural networks, and the terms can in practice be used interchangeably.

between input variables to be determined by the modelling (Goodfellow et al. 2016). As the input data progresses through the hidden layers variables are combined and recombined into newer factors weighted on influence from the prior layer. Thus, for example, a simple economic forecasting model fed with the input variables of GDP growth, unemployment rate, exchange rates, government budget deficit, might then recombine in hidden nodes into new factors. A random example at the first layer of one such recombination (performed by the deep learning model, not by the researcher) might be 0.2*GDP growth + 0.4*exchange rates. As each hidden layer progresses these recombinations could become more and more abstract if that helps with the task set the model.

The addition of hidden layers between input and output is what creates the perceived problem with deep learning—that the process is a 'black box'—in that it is not always clear how inputs have been recombined to create a predicted output. This has obvious implications for use in risk management, where the very presence of a black box at the centre of decision-making can be its own source of risk in a firm.

3.3   Machine Learning and AI Applications for Risk Management

In this section, we provide details and analysis of actual applications of AI and machine learning to various areas of risk management. We categorise risk management using common distinctions in financial risk management, namely: credit risk, market risk, operational risk, and add a fourth category around the issue of compliance.

3.3.1   Application to Credit Risk

Credit risk is economic loss that emanates from the failure of a counterparty to fulfil its contractual obligations (e.g., timely payment of interest or principal), or from the increased risk of default during the term of the transaction. Traditionally, financial firms have employed classical linear, logit, and probit regressions to model credit risk (Altman 1968). However, there is now an increased interest by institutions in using AI and machine learning techniques to enhance credit risk management practices, partially due to evidence of incompleteness in traditional techniques. The evidence is that credit risk management capabilities can be significantly improved through leveraging AI and machine learning techniques due to its ability of semantic understanding of unstructured data.

The use of AI and machine learning techniques to model credit risk is not a new phenomenon though it is a growing one. Back in 1994, Altman and colleagues performed a first comparative analysis between traditional statistical methods of distress and bankruptcy prediction and an alternative neural network algorithm, and concluded that a combined approach of the two improved accuracy significantly (Altman et al. 1994).

It is particularly the increased complexity of assessing credit risk that has opened the door to machine learning. This is evident in the growing credit default swap (CDS) market where there is a lot of uncertain elements involving determining both the likelihood of an event of default (credit event) and estimating the cost of default in case a default takes place. Son et al. (2016) use daily CDS of different maturities and different rating groups from January 2001 to February 2014 to show that nonparametric machine learning models involving deep learning outperform traditional benchmark models in terms of prediction accuracy as well as in proposing practical hedging measures.

The areas of consumer lending and SME lending involve large amounts of potential data and are increasingly relying on machine learning to make better lending decisions. The example of ZestFinance in the opening section is an example of this, but there are a wide number of similar firms operating in this space. There is substantial empirical support for the effectiveness of machine learning. In consumer lending, Khandani et al. (2010) propose a machine-learning technique based on decision trees and SVM that, when tested on actual lending data lead to cost savings of up to 25%. More recently, Figini et al. (2017) show that a multivariate outlier detection machine learning technique improves credit risk estimation for SME lending using data from UniCredit Bank.

3.3.2 *Application to Market Risk*

Market risk is the risk that emanates from investing, trading, and generally from having exposure to financial markets. Kumar (2018) provides a structural overview of how machine learning can help in market risk management, noting the benefits at each stage from data preparation, to modelling, stress testing, and providing a validation trail for model explanation (see also Financial Stability Board 2017).

Trading in financial markets inherently involves the risk that the model being used for trading is false, incomplete, or is no longer valid. This area is generally known as model risk management. Machine learning is particularly suited to stress testing market models to determine inadvertent or emerging risk in trading behaviour. Woodall (2017) describes a variety of current use cases of machine learning for model validation, including the French investment firm Nataxis which at the time of writing was running over 3 million simulations a night using unsupervised learning to establish new patterns of connection between assets and investigating further any simulations that emerged from the testing that showed 'errant' patterns compared to average estimates. Woodall also notes how Nomura uses machine learning to monitor trading within the firm to verify that unsuitable assets are not being used in trading models. An interesting current application of model risk management is the firm *yields.io* which provides real-time model monitoring, model testing for deviations, and model validation, all driven by AI and machine learning techniques.

Another area of focus within the category of market risk of importance to large trading firms is understanding the impact of their trading on market pricing. Day (2017) explores how large trading firms are using AI, and particularly clustering techniques, to avoid the costs of trying to trade into and out of large positions in illiquid markets. He provides a quote from Capital Fund Management, one of the largest hedge funds in France with $11 billion under management, claiming that up to two-thirds of their profit from trades can be lost due to market impact costs. Machine learning techniques significantly help with this issue by identifying connections between assets that are not easily observable and thus allow entering desired positions through a series of related assets rather than taking a large position in a single asset. Cluster analysis particularly helps in this regard (Calvalcante et al. 2016), as can deep learning models (Heaton et al. 2017).

One future direction is to move more towards reinforcement learning, where market trading algorithms are embedded with an ability to learn from market reactions to trades and thus adapt future trading to take account of how their trading will impact on market prices (Hendricks and Wilcox 2014). Another interesting direction is proposed in Chandrinos et al. (2018), based on tests using foreign exchange market trading data, where a combination of neural networks and decision tree techniques are used to provide real-time warnings to traders of changes in underlying trading patterns while trading. Wu and Olson (2015, Chapter 5) also examine the use of machine learning to provide warning signals to traders and demonstrate the benefits of SVM as a suitable technique.

3.3.3   Application to Operational Risk

Operational risk management entails the firm seeking to identify the risk of direct or indirect financial loss emanating from a host of potential operational breakdowns (Moosa 2007). These risks can be internal to the institutions (e.g., inadequate or failed internal processes, people, and systems), or from external events (e.g., frauds, vulnerable computer systems, a failure in controls, operational error, a procedure that has been neglected, or a natural disaster). Given the increase in quantity, variety, and complexity, of operational risk exposures, especially for financial firms, this has introduced a path towards AI and machine learning based solutions (Choi et al. 2017).

AI can assist institutions at various stages in the risk management process ranging from identifying risk exposure, measuring, estimating, and assessing its effects (Sanford and Moosa 2015). It can also help in opting for an appropriate risk mitigation strategy and finding instruments that can facilitate shifting or trading risk. Thus, use of AI techniques for operational risk management, which started with trying to prevent external losses such as credit card frauds, is now expanding to new areas involving the analysis of extensive document collections and the performance of repetitive processes, as well as the detection of money laundering that requires analysis of large datasets.

The detection of financial fraud is another commonly referenced risk management use case for machine learning and AI. Here, banks attempt to control financial fraud through evaluating the best ways to protect their systems, their data, and ultimately their clients. AI's ability to introduce better process automation can accelerate the pace of routine tasks, minimise human error, process unstructured data to screen out relevant content or negative news, and determine individuals' connectedness to evaluate risky clients and networks. This same network analysis can also be used to monitor employees and traders. Clustering and classification techniques can be used to establish behaviour-based trader profiles, where combinations of trade data, electronic and voice communications records enable banks to observe emerging patterns of behaviour to predict latent risks and detect links between employees. It also enables banks to generate and prioritise alerts based on types of suspicious activity and the level of risks involved. Ngai et al. (2011) provide an excellent overview of the core AI techniques used in the financial fraud detection, and note the main techniques applied as being decision trees and neural networks.

As a practical application, five of the biggest banks in the Nordics have recently joined together to establish a joint anti-money laundering infrastructure—known as the Nordic KYC Utility. An AI-based infrastructure will help comply with regulations and requirements related to KYC (Know Your Customer) regulations and avoid the imposition of fines by regulators. Similarly, HSBC is introducing AI technology, developed by data analytics firm Quantexa, to monitor their anti-money laundering processes. There are practical efforts in fraud prevention too. For example, a joint venture of Royal Bank of Scotland and Vocalink in the UK is creating a machine learning system to scan transactions by small and large business customers to identify and circumvent false invoices and potential instances of fraud. A study by Colladon and Remondi (2017) using real data from 33,000 transactions of an Italian factoring firm shows the effectiveness of such analysis in fraud detection (see also Demetis 2018).

3.3.4   Application to RegTech

Compliance with risk management regulations is a vital function for financial firms, especially post the financial crisis. While risk management professionals often seek to draw a line between what they do and the often-bureaucratic necessity of regulatory compliance, the two are inextricably linked as they both relate to the overall firm systems for managing risk. To that extent compliance is perhaps best linked to enterprise risk management, although it touches specifically on each of the risk functions of credit, market, and operational risk.

RegTech, an analogous area to FinTech focused on compliance discussed further in Chapter 6, has thus arisen to assist firms in the increasing demands of meeting compliance. In this area AI and machine learning is starting to play a significant role driven by the sheer volume of data that needs to be assessed as well as the non-conventional nature of this data. While much of the role of AI and machine learning in RegTech relates to topics already discussed in prior sections, the key advantage of machine learning in a pure RegTech sense is the ability for continuous monitoring of firm activities. Arner et al. (2016) note this ability for real-time insights and therefore avoiding compliance breaches rather than having to deal with the consequences of breaches after they have occurred. Other advantages noted are the ability to free up regulatory capital due to the better monitoring, as well as automation reducing some of the estimated $70 billion that major financial institutions spend on compliance each year.

A key player in this field is IBM following their acqui-hire purchase of Promontory (a 600-staff RegTech startup), and they now offer a range of AI-driven solutions for reducing RegTech costs, demonstrating the widespread industry interest in the area not just confined to startups. For example, real-time voice conversation analysis to being used to ensure compliance through a combination of IBM's Watson AI expertise and Promontory's domain-specific expertise. This involves translating voice conversations to text and then classifying this text using natural language processing into categories that identifies potential non-compliance. Other applications of machine learning include the automatic reading and interpretation of the implications of regulatory documentation, again using natural language processing, as a London-based Waymark currently offers to financial institutions.

3.4   The Challenges and Future of Machine Learning and AI for Risk Management

There are some significant practical issues that need to be addressed before AI and machine learning techniques for risk management can claim its full potential. The most important of these is the availability of suitable data. Although machine learning packages for Python and R can easily read all types of data from Excel to SQL and can perform natural language processing and process images, the speed with which machine learning solutions have been proposed has not kept pace with firms' abilities to suitably organise the internal data they have access to. Data is often held in separate silos across departments, perhaps on different systems, and perhaps with internal political and regulatory issues restricting the sharing of data. Important data might not even be recorded as data but rather kept as informal knowledge of the firm.

Another issue is the availability of skilled staff to implement these new techniques. A survey of the top 1000 firms in the United States on AI implementation in their firms found that their biggest concern in the implementation of AI was the readiness and ability of staff to understand and work with these new solutions (Wilson et al. 2017). Training a skilled cohort of staff is something that will take time, although Goldman Sachs, among other firms, has recently attempted to bypass this problem by developing a campus with space for 7000 workers in India where prevalence of these skills is more present.

There are also practical issues over how accurate machine learning solutions actually are. The range of testing approaches available within machine learning is growing rapidly, and that is a good thing, but it is also driven by the evident limitations of the previous methods and the need to overcome those limitations. This suggests that firms cannot simply 'apply' a machine learning risk management solution, but rather that it is a continuous process requiring a constant evaluation of whether their particular machine learning solution is currently considered best practice. When it comes to AI, where there is some or full automation of process from data gathering to decision-making, the need for human oversight will become even more pressing. The case of Knight Capital in 2012 serves to illustrate this risk, with their stock trading automation through algorithms resulting in a loss of $440 million in the space of just 45 minutes.[3] As AI starts to automate lending and credit risk decisions it will be imperative to ensure that such risks can be controlled before handing over control.

This last point serves to introduce the final major issue around transparency and ethics which AI-driven solutions need to further address. Transparency is especially an issue for the increasingly popular deep learning method of machine learning as the models work on hidden layers between the input data and the output decision. A black box system of this type of not conducive to effective risk oversight and can cause regulatory compliance issues especially around demonstrating model validity. A more hypothetical issue related to this is that models used by different firms might converge on similar optimums for trading causing systematic risk as well as loss to the firm. There are also broader ethical issues. For example, discrimination against race in lending decisions is

---

[3] http://fortune.com/2012/08/02/why-knight-lost-440-million-in-45-minutes/.

widely legislated against, as are increasingly discrimination based on gender and sexuality. Normally these restrictions are incorporated as hard-coded rules in AI and machine learning techniques concerning credit risk and lending decisions. However as deep learning becomes more popular it becomes much harder to police that the model is not inadvertently making decisions based on indirect proxies for these red lines. This is especially the case as more and more atypical data is incorporated in risk management. These ethical aspects are only likely to increase in significance over time.

Leaving these important issues aside for now, it is worth considering the future role for AI and machine learning in risk management from a more positive angle. One obvious conclusion is that the time-consuming and costly nature of risk management will diminish significantly. For example, BBVA, the second largest bank in Spain, has 8000 of their 137,000 staff working in compliance. They are investing heavily in AI solutions to reduce this compliance cost base. The ability of AI and machine learning to automate repetitive tasks and to organise, retrieve, and cluster non-conventional data such as documents is naturally going to confer cost benefits on firms that move more into this area.

AI will also increasingly deliver accurate real-time information on all types of risks being taken by the firm. As data organisation becomes more orientated towards use by AI, real-time advice will become a pervasive presence. The following step from real-time knowledge of risks being taken is pre-emptive notice of risks. To some extent this is the holy grail of an AI-driven risk management system—to be able to accurately know in advance firm risks, be they market, operational, or credit risk. The techniques of machine learning offer this ability in the way that traditional statistical techniques could never hope to. Thinking even further ahead there is no technological impediment to a truly AI risk management system that will automatically intervene to prevent unwarranted risks, to immediately unwind dangerous exposures, and to dynamically adjust the risk appetite of the firm based on the system's estimate of the broader risk environment. Although that then will present its own risks which will in turn have to be managed, keeping risk management professionals happily employed (albeit in a fast-changing environment) for the foreseeable future.

We thus conclude on a positive note (albeit with some issues to consider), about how machine learning and AI is transforming the way we carry out risk management. The issue for the established risk management functions in organisations to now consider is if they wish to avail of these opportunities, or if instead it will fall to current and new FinTech firms to seize this space.

REFERENCES

Altman, E. I. (1968). Financial ratios, discriminant analysis and the prediction of corporate bankruptcy. *The Journal of Finance, 23*(4), 589–609.

Altman, E. I., Marco, G., & Varetto, F. (1994). Corporate distress diagnosis: Comparisons using linear discriminant analysis and neural networks (the Italian experience). *Journal of Banking & Finance, 18*(3), 505–529.

Arner, D. W., Barberis, J. N., & Buckley, R. P. (2016). The emergence of RegTech 2.0: From know your customer to know your data. *Journal of Financial Transformation, 79* (UNSW Law Research Paper No. 17–63). Available at SSRN: https://ssrn.com/abstract=3044280.

Bostrom, N. (2014). *Superintelligence: Paths, dangers, strategies.* Oxford: Oxford University Press.

Cavalcante, R. C., Brasileiro, R. C., Souza, V. L., Nobrega, J. P., & Oliveira, A. L. (2016). Computational intelligence and financial markets: A survey and future directions. *Expert Systems with Applications, 55,* 194–211.

Chandrinos, S. K., Sakkas, G., & Lagaros, N. D. (2018). AIRMS: A risk management tool using machine learning. *Expert Systems with Applications, 105,* 34–48.

Choi, T. M., Chan, H. K., & Yue, X. (2017). Recent development in big data analytics for business operations and risk management. *IEEE Transactions on Cybernetics, 47*(1), 81–92.

Colladon, A. F., & Remondi, E. (2017). Using social network analysis to prevent money laundering. *Expert Systems with Applications, 67,* 49–58.

Day, S. (2017). Quants turn to machine learning to model market impact. Risk.net. Available at: https://www.risk.net/asset-management/4644191/quants-turn-to-machine-learning-to-model-market-impact. Last accessed 17 August 2018.

Demetis, D. S. (2018). Fighting money laundering with technology: A case study of Bank X in the UK. *Decision Support Systems, 105,* 96–107.

Figini, S., Bonelli, F., & Giovannini, E. (2017). Solvency prediction for small and medium enterprises in banking. *Decision Support Systems, 102,* 91–97.

Financial Stability Board. (2017). Artificial intelligence and machine learning in financial services. Available at: http://www.fsb.org/wp-content/uploads/P011117.pdf. Last accessed 17 August 2018.

Goodfellow, I., Bengio, Y., Courville, A., & Bengio, Y. (2016). *Deep learning* (Vol. 1). Cambridge: MIT Press.

Heaton, J. B., Polson, N. G., & Witte, J. H. (2017). Deep learning for finance: Deep portfolios. *Applied Stochastic Models in Business and Industry, 33*(1), 3–12.

Hendricks, D., & Wilcox, D. (2014). A reinforcement learning extension to the Almgren-Chriss framework for optimal trade execution. In *IEEE Conference on Computational Intelligence for Financial Engineering & Economics (CIFEr)* (pp. 457–464).

Hu, Y., Zhang, X., Feng, B., Xie, K., & Liu, M. (2015). iTrade: A mobile data-driven stock trading system with concept drift adaptation. *International Journal of Data Warehousing and Mining (IJDWM), 11*(1), 66–83.

Khandani, A. E., Kim, A. J., & Lo, A. W. (2010). Consumer credit-risk models via machine-learning algorithms. *Journal of Banking & Finance, 34*(11), 2767–2787.

Kumar, P. P. (2018). Machine learning for model development in market risk. *GARP Institute*. Available at: https://www.garp.org/#!/risk-intelligence/all/all/a1Z1W000003fM0yUAE?utm_medium=social&utm_source=facebook&utm_content=org_whitepaper&utm_term=machinelearning&utm_campaign=sm_riskintelligence. Last accessed 17 August 2018.

Moosa, I. A. (2007). *Operational risk management*. New York: Palgrave Macmillan.

Nazemi, A., Heidenreich, K., & Fabozzi, F. J. (2018). Improving corporate bond recovery rate prediction using multi-factor support vector regressions. *European Journal of Operational Research*, forthcoming.

Ngai, E. W., Hu, Y., Wong, Y. H., Chen, Y., & Sun, X. (2011). The application of data mining techniques in financial fraud detection: A classification framework and an academic review of literature. *Decision Support Systems, 50*(3), 559–569.

Sanford, A., & Moosa, I. (2015). Operational risk modelling and organizational learning in structured finance operations: A Bayesian network approach. *Journal of the Operational Research Society, 66*(1), 86–115.

Shieber, S. M. (Ed.). (2004). *The turing test: Verbal behavior as the hallmark of intelligence*. Cambridge: MIT Press.

Son, Y., Byun, H., & Lee, J. (2016). Nonparametric machine learning models for predicting the credit default swaps: An empirical study. *Expert Systems with Applications, 58*, 210–220.

van Liebergen, B. (2017). Machine learning: A revolution in risk management and compliance? *Journal of Financial Transformation, 45*, 60–67.

Wilson, H. J., Daugherty, P., & Bianzino, N. (2017). The jobs that artificial intelligence will create. *MIT Sloan Management Review, 58*(4), 14–16.

Woodall, L. (2017). Model risk managers eye benefits of machine learning. Risk.net. Available at: https://www.risk.net/risk-management/4646956/model-risk-managers-eye-benefits-of-machine-learning. Last accessed 17 August 2018.

Wu, D. D., & Olson, D. L. (2015). *Enterprise risk management in finance*. New York: Springer.

Zhong, X., & Enke, D. (2017). Forecasting daily stock market return using dimensionality reduction. *Expert Systems with Applications, 67,* 126–139.

CHAPTER 4

What FinTech Can Learn from High-Frequency Trading: Economic Consequences, Open Issues and Future of Corporate Disclosure

Eleonora Monaco

Abstract This chapter provides a review on key literature on High-Frequency Trading (HFT) over an 11-year period. Using a thematic analysis, the main themes developed within this research stream are identified and insights on the evolution of theory in relation to HFT are presented. This analysis highlights that the effects of HFT on market liquidity, trading strategies and speed, implications for market structure changes, and the relationship between the "scriptability" of corporate disclosure and HFT short-term information advantage, are key themes.

E. Monaco (✉)
Católica Porto Business School,
Universidade Católica Portuguesa, Porto, Portugal
e-mail: emonaco@porto.ucp.pt

E. Monaco
European Capital Markets Cooperative Research Centre,
Chieti, Italy

© The Author(s) 2019
T. Lynn et al. (eds.), *Disrupting Finance*, Palgrave Studies
in Digital Business & Enabling Technologies,
https://doi.org/10.1007/978-3-030-02330-0_4

The analysis also suggests that many open questions remain unanswered including more recent HFT trading strategies and complex techniques applied to analyse the content of both voluntary and mandatory corporate disclosure. As capital markets evolve, HFT's speed may no longer be sufficient to maintain competitiveness. The chapter concludes with a discussion of future trends and areas for research on HFT.

Keywords High-frequency trading · Literary review · Market quality · Regulation · Corporate disclosure scriptability

4.1   Introduction

In the last years, investment in financial technology (FinTech) grew by 201% around the world; total venture capital investments only grew by 63% in the same period (Aldridge and Krawciw 2017). According to Informilo.com, in 2015 payment services, online loans, data analytics and automated investing have resulted in the fastest-growing areas for big data in finance. Specifically, investment automation and other new related technologies have transformed the structure of capital markets. Reducing market-wide latency, the introduction of both co-location services and fast trading platforms enable new sophisticated investors to enter into the market. Therefore, using high speed and high-performance computing, sophisticated tools and algorithms, algorithmic traders (AT) rapidly trade securities in the main stock exchanges around the world. These changes and the behaviour of market participants attract considerable attention by both the academic community and policymakers. Many papers discuss the role played by AT in capital markets as well as their trading strategies and consequences for market quality. Similarly, market regulators have expressed concerns about the growing participation of ATs and the costs associated with monitoring their activities.

This chapter provides a review of the High-frequency trading (HFT) literature based on 11 years of publications, discusses HFT consequences on capital markets, and suggests future research directions. Following a similar approach adopted by prior studies (Massaro et al. 2016), this chapter aims to answer three questions:

- What are the major themes that have been discussed in HFT research field?

- What are the main issues and critique on HFT activity?
- What is the future of HFT research?

The remainder of this chapter is organised as follows. The next section considers the differing approaches to defining HFT by both regulators and scholars and presents an overview of common datasets used to investigate HFT activity. This is followed by a summary of the methodology used for the literature review and associated data collection. The results of the literature review are then presented. The chapter concludes with a discussion of the findings and directions for future research.

4.2   HIGH-FREQUENCY TRADING: DEFINITION AND DATA

In general, total trading activity can be classified into two main categories: algorithmic trading (AT) and non-algorithmic trading activity (NAT) depending on whether or not market participants use algorithms to make trading decisions without human intervention (ESMA 2014). The European Markets in Financial Instruments Directive (MiFID II) provides two different definitions of the concepts HFT and AT, where the former is a subset of the latter. Specifically, AT is defined as "[...] trading in financial instruments where a computer algorithm automatically determines individual parameters of orders, such as whether to initiate of an order or how to manage the order after its submission, with limited or no human interaction" (MiFID II 2014, p. 384). This does not include any system that is only used for the purpose of routing orders to one or more trading venues or for the processing of orders involving no determination of any trading parameters or for the confirmation of orders or the post-trade processing of executed transactions (MiFID II 2014). Moreover, the MiFID II (2014, pp. 384–385) defines HFT as "an algorithmic trading technique that is characterized by an infrastructure that minimize network and other type of latencies using specific facilities as co-location, proximity hosting or high-speed direct electronic access and by a system determination of order initiation, generation and execution without human intervention for trades or orders".

The U.S. Securities and Exchange Commission takes a broader approach in defining HFT as "professional traders acting in a proprietary capacity that engage in strategies that generate a large number of trades on a daily basis" (SEC 2010, p. 45). Similarly, the Australian Securities and Investments Commission emphasises HFT's ability to generate large numbers of orders, many of which are cancelled rapidly and to hold positions for very short time horizons (ASIC 2010).

Although there is no common definition of HFT, several regulatory agencies and scholars do attempt to identify two main features and trading strategies of HFT, namely: (i) the automation of the trading process, and (ii) the high speed of transactions and submission (cancellation) of orders.

Different methods have been applied to classify HFT activities. For example, the definition used by SEC (2010) allows the identification of HFT activities but fails to detect large blocks of HFT. Some scholars instead detect such blocks by focusing on the evidence of high trading volume and balance inventory (Kirilenko et al. 2017) or on complex trading strategies characterised by "series of submissions, cancellations, and executions that are linked by direction, size and timing, and which are likely to arise from a single algorithm" (Hasbrouck and Saar 2013, p. 656).

Given that only a few datasets (such as E-Mini Data and NASDAQ data) allow the identification of HFT, most studies are based on proxies to detect HFT activity and highlight the effects of HFT on capital markets. The current HFT datasets available can be classified into five categories:

- Data for equity trading on NASDAQ;
- Data on trading in the E-Mini;
- Data that CFTC and SEC staff used to prepare their report on the market disruption that occurred in 2010 (Flash Crash);
- A variety of datasets made available to researchers by exchanges and regulators that require proxies to identify HFT activity.

According to Boehmer et al. (2018) and Hendershott et al. (2011), message traffic can include new order submissions, modifications or order cancellations. Hence, the main proxies used by researchers are trading volume (Clark-Joseph 2013; Baron et al. 2017; Kirilenko et al. 2017), raw messages, the ratio of both trading volume and number of messages (Hendershott et al. 2011), and the ratio of messages and total transactions.

4.2.1   *Methodology*

This section explains the methods for selecting and reviewing the articles examined in this study. Following the methodology used in Massaro et al. (2016) and similar approaches used by prior studies, a dataset of articles was constructed. The dataset counts HFT articles published in the main accounting and finance journals featured in the Scopus database by Elsevier for the 11-year period from 1 January 2007 to 15 May 2018. In order to be included in the sample, articles must mention the terms "high-frequency trading" or "algorithmic trading" in the title, abstract or keywords. Those papers that did not meet the required conditions were discarded, reducing the list of articles to 265 articles (either published or forthcoming on 15 May 2018). Author details, article title, year of publication, SCOPUS citations, affiliation of authors and location were collected.

Firstly, the articles were classified based on whether they were published in generalist or specialist journals. 87 articles were published in general journals, while 178 articles were published in specialist journals. The latter category includes journals whose scope focuses specifically on HFT activity e.g. Algorithmic Finance Journal, on trading issues and structures of financial markets e.g. *Journal of Finance, Journal of Financial Economics, Journal of Financial Markets, Journal of Empirical Finance* and others.

Secondly, the selected articles were classified by examining their citations to measure the academic impact of each article and to provide insights into the evolution of the literature (Table 8.1). The journals were further categorised by ABS Journal Ranking (or not) and the main topics covered in each article were identified.

4.2.2   Descriptive Statistics

This section reports some descriptive statistics to provide a clear picture of the evolution of the HFT literature. The first article that refers to the "activity of algorithmic trading" was published in 2007 (Prix et al. 2007) and describes the systematic patterns in the submission and cancellation of certain Xetra orders.[1] Figure 4.1 shows the evolution of the volume of papers published in generalist or in specialist journals ranked by Scopus in the chose time period. A sharp increase in the articles is evident from 2010, following the "Flash Crash" in May of that year, which signalled the start of increased academic discussion of the consequences of HFT activity on capital markets.

However, the largest number of contributions was published in the four years between 2013 and 2017 in which not only specialist journals but also generalist journals published articles (123 and 58 articles,

---

[1] The advantage of "the predatory traders" over uninformed traders has been discussed for the first time by Brunnermeier and Pedersen (2005).

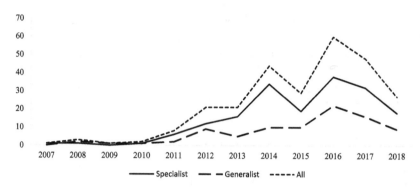

Fig. 4.1 Volume of articles on HFT ranked by Scopus in the period 2007–2018

respectively) noting the consequences of HFT activity on capital markets as well as the impact of new regulation, released by different countries, to limit their presence.

An examination of author affiliation allows the identification of the institutions in which researchers conducted their studies on HFT. This analysis suggests that the leading institutions publishing in the field of AT research include The University of Toronto, The University of California Berkeley, The University of Oxford, The University of Sydney, The University of Washington and Imperial College London. The majority of studies were developed in USA (30%), UK (14%), France (7%) and Australia (7%) (Fig. 4.2).

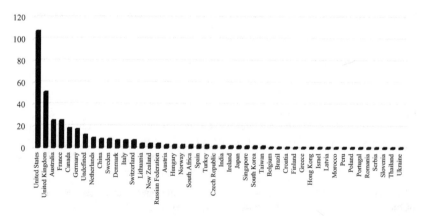

Fig. 4.2 Absolute frequency of articles by Scholars' location on the final sample

Analysis of the research quality of HFT publications suggests that since the appearance of the seminal paper of Hendershott et al. (2011), 21 articles were published in ABS 4* Journals (*Journal of Finance, Journal of Accounting Research, Journal of Accounting and Economics, Journal of Financial Economics, Quarterly Journal of Economics, Review of Economic Studies, Review of Financial Studies*), 5 articles were published by ABS 4 journals and another 103 articles by ABS 3 journals (Tables 4.1 and 4.2) AT research is not only topical but is considered a priority by the major high-quality journals.

4.3   RESULTS

4.3.1   Thematic Analysis

A thematic analysis was performed to identify and classify the main themes discussed in the literature (Clarke and Braun 2013). Furthermore, a citation analysis based on the Scopus index (Dumay 2014) was used to identify articles and authors that have the most impact in HFT research (Garfield 1977). Table 4.3 presents the distribution of a subsample of 168 articles that have been classified by topic. Only the articles published on the highest rated ABS-ranked journals 4*, 4, 3 and 2 (with at least 1 citation) have been included in the subsample. The table reports both frequency and sum of citations by topic. The latter table suggests four main research paths i.e.

1. effects on market quality and price discovery,
2. trading strategies,
3. impact of financial markets structure, regulation and co-location, and
4. HFT reaction to corporate disclosure.

The most cited articles discuss both about effects of HFT on market quality (1098 citations related to the 34.5% of articles in the sample) and the trading strategies implemented (259 citations related to 36 articles).

4.3.2   Impact of HFT

4.3.2.1   Effects on Market Quality
In the last years, several studies have examined the consequences of HFTs on market quality by investigating both various dimensions of

Table 4.1 Distribution of articles included in Scopus database and in ABS ranked journals

Year	Articles	ABS ranking					Other	Specialist	Specialist (%)	Generalist	Generalist (%)
		4*	4	3	2	1					
2007	1			1				1	0.38		0.00
2008	3			2			1	1	0.38	2	0.75
2009	1						1		0.00	1	0.38
2010	2			1	1			1	0.38	1	0.38
2011	8	1		5		1	1	6	2.26	2	0.75
2012	21			8	2	3	8	12	4.53	9	3.40
2013	21	2	2	9	2	2	4	16	6.04	5	1.89
2014	44	4		23	4	4	9	34	12.83	10	3.77
2015	29	4		10	7	3	5	19	7.17	10	3.77
2016	60	2	2	20	12	5	19	38	14.34	22	8.30
2017	48	6	1	11	22	4	4	32	12.08	16	6.04
2018	27	2		13	8	2	2	18	6.79	9	3.40
Total	265	21	5	103	58	24	54	178	67.17	87	32.83

Table 4.2 Distribution of articles included in Scopus database and in the top ABS ranked journals (4*−4−3)

Journals	N	%	ABS ranking
Quantitative Finance	19	7.17	3
Journal of Financial Markets	17	6.42	3
Financial Review	12	4.53	3
Journal of Financial Economics	9	3.40	4*
Journal of Banking and Finance	9	3.40	3
Journal of Futures Markets	7	2.64	3
Journal of Finance	5	1.89	4*
Journal of International Financial Markets, Institutions and Money	5	1.89	3
Annual Review of Financial Economics	3	1.13	3
European Journal of Finance	3	1.13	3
International Review of Financial Analysis	3	1.13	3
Journal of Accounting Research	2	0.75	4*
Review of Financial Studies	2	0.75	4*
Journal of Financial and Quantitative Analysis	2	0.75	4
Applied Econometrics	2	0.75	3
European Financial Management	2	0.75	3
Finance and Stochastics	2	0.75	3
Journal of Business Ethics	2	0.75	3
Journal of Economic Behavior and Organization	2	0.75	3
Journal of Economic Dynamics and Control	2	0.75	3
Journal of Empirical Finance	2	0.75	3
Journal of Financial Econometrics	2	0.75	3
Mathematical Finance	2	0.75	3
Review of Quantitative Finance and Accounting	2	0.75	3
Journal of Accounting and Economics	1	0.38	4*
Quarterly Journal of Economics	1	0.38	4*
Review of Economic Studies	1	0.38	4*
Business Ethics Quarterly	1	0.38	4
Journal of Economic Perspectives	1	0.38	4
Journal of Economic Theory	1	0.38	4
Financial Analysts Journal	1	0.38	3
Financial Markets, Institutions and Instruments	1	0.38	3
International Journal of Finance and Economics	1	0.38	3
Journal of Financial Services Research	1	0.38	3
Journal of International Money and Finance	1	0.38	3
Articles in ABS 4*, 4, 3 ranked journals	129	48.68	–
Articles in ABS 2,1 ranked journals and other journals	136	51.32	–
Total articles	265	100	

Table 4.3 Absolute and percentage frequency of articles by topic and by Scopus citations

Topics	Frequency	%	Citations	%
Market quality	58	34.5	1098	91.3
Trading strategies and speed	36	21.4	259	21.5
Financial markets structure	14	8.33	134	11.1
Regulation and co-location	11	6.15	74	6.16
Price discovery	10	5.95	113	9.4
Flash crash	5	2.98	42	3.49
Financial disclosure	3	1.79	46	3.83
Transaction costs	3	1.79	26	2.16
Dark market fragmentation	2	1.19	11	0.92
Investors strategies	2	1.19	39	3.24
Order-to-trade	3	1.79	9	0.75
Disposition effect	1	0.6	6	0.5
Hawkes processes	1	0.6	6	0.5
Other	19	11.3	238	19.8
Total	168	100	1202	100

price discovery, short-term volatility and stock liquidity (Hasbrouck and Saar 2013; Malinova et al. 2013; Menkveld 2013; Conrad et al. 2015). Specifically, researchers found that the introduction of HFT has been accompanied by a reduction in trading costs (Angel et al. 2015; Jones 2013) and by an improvement in price efficiency (Carrion 2013; Brogaard et al. 2014; Chaboud et al. 2014).

Examining the NYSE automated quote dissemination in 2003, the seminal paper of Hendershott et al. (2011) measures the causal effect of AT on liquidity, and demonstrates that AT activity narrows spreads, reduces both adverse selection and trade-related price discovery. Other studies highlight how HFT's market share has boomed over the number of years. Examining the role of AT in liquidity supply in 30 DAX stocks on the Deutsche Boerse, Hendershott et al. (2011) find that AT represent 52% of market order volume and 64% of nonmarketable limit order volume. Similarly, Hagströmer and Nordén (2013) report that market markets constitute the lion share of HFT trading volume (63–72%) and limit order traffic 81–86% of NASDAQ-OMS Stockholm Exchange. These results demonstrate that AT consume liquidity when it is cheap (narrow bid-ask spread or other proxies such as effective spread, percentage spread or higher depth) given that it is less likely to submit new orders, to cancel their orders and more likely to initiate trades. Similarly,

Yao and Ye (2018) find that HFT liquidity supply is larger for stocks for which the spread is constrained to be large because of tick size. Jarnecic and Snape (2014) describe the HFT provision of liquidity on an on-going basis which robust to fast versus slow and volatile markets resolving in this way the temporal liquidity imbalances.

Other studies examine HFT consequences around specific events such as earnings announcements (EPMs) (Frino et al. 2017), short sale-ban (Brogaard et al. 2017), predictable trades (Bessembinder et al. 2016) and extreme price movements (EPMs) (Brogaard et al. 2018), generally confirming the liquidity improvement.

Using the September 2008 short sale-ban, Brogaard et al. (2017) disentangle the separate impact of short selling by HFTs and non-HFTs. They suggest that non-HFTs increase liquidity (as measured by bid-ask spreads) while HFTs' short selling has the opposite effect by adversely selecting limit orders which decrease liquidity during extremely volatile short-sale ban period. Brogaard et al. (2018) investigate the activity of a common endogenous liquidity providers (ELPs), such as HFTs, around EPMs discovering that on average HFTs provide liquidity not only during normal market conditions but also around a market stress such as EPMs.

Other studies have focused on HFT strategies and their influence on market quality. For example, Hagstromer and Nordén (2013) examine the tick-size changes and find that the activity of market-making HFTs mitigates intraday price volatility (or short-term volatility for Boehmer et al. 2015), and thus can contain the deterioration of market quality.

4.3.2.2   HFT's Trading Strategies and Speed

The SEC's Concept release on Equity Market Structure recognised not only that the advent of HFTs is one of the most significant market structure developments in recent years (SEC 2010) but identified the existence of four types of short-term HFTs trading strategies—passive market making, arbitrage, structural and directional.

"Passive market making" involves the submission of non-marketable resting orders that provide liquidity to the marketplace at specified prices. Following this strategy, HFT orders are not executed immediately but rest on an order book and prices are updated frequently to reflect market conditions. In this way HFTs generate a great number of order cancellations or modifications as orders are updated and earn a spread between bids and offers. This passive strategy decreases effective spread as demonstrated by Menkveld (2013). Similarly, the "arbitrage strategy" does not depend on directional price movements but on price convergence seeking to keep the differences between related products or markets.

If HFTs follow "structural strategies", they attempt to find the weakness in the market structure to take advantage of the other participants. In fact, given HFT can access market data in real time, the lower latency allows them to establish prices both on the seller and buyer side. In contrast, "directional strategies" involve establishing a short position in anticipation of a price move up or down. The Concept Release (SEC 2010) requested comment on two types of directional strategies, order anticipation and momentum ignition, that "may pose particular problems for long-term investors" and "may present serious problems in today's market structure".

Regarding HFT speed, Menkveld (2016) estimates that algorithms have a response time in the order of microseconds (one microsecond is 10^{-6}). Therefore, even if HFT effects on the market are known, both in terms of less adverse-selection cost, tighter bid-ask spread (higher liquidity), frequent quote updates between trades and higher price discovery between the quoted updates, and a higher trade probability (Brogaard et al. 2018), the main concerns of regulators remain whether their speed represents a barrier that limits retail investor trading activity. Moreover, regarding the open question regarding whether HFTs are better informed agents, few studies (Goettler et al. 2009; Aït-Sahalia and Saglam 2013) find a liquidity improvement when market makers become more informed about fundamental value. Agents arrive randomly and, conditional on the state of the limit-order book, they can choose to send either a limit or a market order. Other studies discussed HFT speed to cancel their outstanding limit order after news (Hoffmann 2014), an endogenous strategy that post limit orders at less aggressive prices, reducing the trade rate.

Finally, observing the price competition in a limit-order market Bongaerts et al. (2016) discover that the increase of HFT has as the final effect a general improvement of liquidity.

4.3.2.3 Market Structure, Co-location and Regulation After the Flash Crash

On 6 May 2010, the US financial markets experienced the Flash Crash. Nearly one trillion US dollars' worth of equity vanished in minutes resulting from a large automated selling program being rapidly executed in the E-mini S&P 500 stock index futures market (Kirilenko et al. 2017).

Although HFTs were blamed for this systemic intraday event, an investigation of the FINRA Dataset shows that 6 out of 12 HFTs reduced their trading activity in the market "sometime after the crash which caused a decline in overall market liquidity. Hence, High-frequency traders did accelerate the rate of crash" (Chung et al. 2016, pp. 17–18).

The Flash Crash highlighted for the first time both the changes in trading speed and in the market's structure, a new arena where the low-frequency traders (LFTs) can only fail to defend themselves from predatory HFTs strategies (Goldstein et al. 2014). It remains still unclear if it is the presence of a weak financial market structure that generates negative events (like the Flash Crash) or whether the latter can be caused by HFT activity. In this respect, Conrad et al. (2015, p. 290) discuss "[...] while dislocations are harmful to market integrity, it is important to recognize that some discontinuities have always occurred in markets (even before the age of electronic trading), just as flickering quotes have existed well before the advent of high-frequency quotation...if liquidity provision is not mandated by law, liquidity providers can always exit without notice, exposing marketable orders to price risk".

Some Scholars argue that exchanges have modified their market structure (Menkveld and Yueshen 2017) to attract more high-frequency traders by, among other things, permitting "algorithmic traders to co-locate their servers in the market's data centre" (Hendershott et al. 2011). This co-location reduces latency and permits HFT to more quickly adjust their quotes as market conditions change and to decrease bid-ask spreads and increase market depth in the period following the introduction of these new facilities (Frino et al. 2014). Using colocation services as a proxy for AT and HFT, Aitken et al. (2014) examine the impact of changes in AT and HFT on trade size across 24 stock exchanges around the world. Mixed results on AT and HFT effects on the average trade size (used to identify AT and HFT start dates) were found. The study also demonstrated, for the first time, that even if the introduction of co-location facilities leads to the presence of HFT, the "colocation dates" do not properly measure effective AT and HFT (Aitken et al. 2014) that may enter into the market a few months before or after the co-location launch (Frino et al. 2017).

While on one hand the main stock exchanges seek to attract a larger number of HFT, both reducing the low latency and introducing new trading platforms, on the other hand many regulators around the world are trying to limit the massive volume of messages (orders), as submissions and cancellations, made by high-frequency traders. More recently, many regulators have attempted to discourage the HFT activity by introducing a specific tax to limit high volumes of messages and cancellations despite the lack of agreement on the negative effects of this legislation on capital market quality. On 1 August 2012, the French government introduced a financial transaction tax applicable on cancelled orders made by high-frequency traders where all orders cancelled or modified within half-second time span are taxed. The tax of 0.01% is applied on modified or cancelled orders of French HFT when OTR is greater than five (Chung et al. 2016), even if in this case it did not have any negative on market quality, both in term of trading volume, volatility, spreads and depth (Colliard and Hoffmann 2017).

Similarly, on 1 March 2013 Italy introduced the iFTT (Italian Financial Transaction Tax) imposing tax on (1) the transfers of the ownership of shares and other participating financial instruments, (2) transactions in derivative financial instruments and other transferable securities, and (3) HFT (MEF 2013). The initiative was launched by the Italian Securities and Exchange Commission (CONSOB) which introduced the iFTT with the specific goal of containing the rapid placement and cancellation of orders. In fact, a recent study suggests that orders' cancellations "can generate a misleading representation of the actual depth of the order book, creating favourable conditions for market manipulation" (Friederich and Payne 2015, p. 215). The iTFF was the first order-to-trade ratio tax (OTR) to attempt to reduce the perceived harmful behaviours of HFT[2] but the new regulation resulted in lower average Italian OTRs (Caivano et al. 2012) and a negative effect on market liquidity (Friederich and Payne 2015).[3] Despite these findings, the taxation of HFT is set to continue. Norway, Germany and Canada have introduced OTR to limit HFT activity (Malinova et al. 2013; Haferkorn and Zimmermann 2014).

4.3.3   HFT Reaction to Corporate Disclosure

Recently Allee et al. (2018) demonstrated the effect of the "scriptability" of firm disclosures on capital markets. Scriptability represents the relative

---

[2] The tax became effective in Borsa Italiana on April 2012 and it intended to charge a fee to HFT with OTF higher than 100:1, 500:1 or 1000:1 (see Grant and Rachel 2012).

[3] Opposite results have been found by Capelle-Blancard (2016).

ease with "which a computer program or a computer programmer can transform the large amounts of unstructured data contained in various firm disclosures into usable information (Bloomfield 2002)". The general assumption here is that "more scriptable" filings allow HFT (and other sophisticated investors) gain a short-term information advantage to react and trade quickly and increase the information asymmetry immediately following disclosures; bid-ask spreads increase by 20–25% in the 30–60 seconds following Form 4 filings (Rogers et al. 2017). Similar results have been found by Frino et al. (2017) that demonstrate that spreads increase at the time of the event and decrease in the following minutes.

4.4   Conclusion and Future Research Directions

The advent of HFT and the introduction of co-location services and other facilities irredeemably changed market structures around the world. Consequently, transaction costs have decreased sharply—by over 50% for both retail and institutional investors (Menkveld 2016). Several studies discuss the consequences of HFT activity on market quality and find a rise in both trading volume and in the number of orders (trades), as well as large increases in the number of submissions (messages) and cancellations. However, the lack of identification codes (in the main financial dataset available) does not allow the disentanglement of trading activity by different type of investors (institutional vs. retail investors). As a result, both the number of orders and the number of submissions or cancellations (messages) are commonly used to build proxies that allow the detection of HFT activity thereby allowing scholars to detect the consequences of HFT activity in the main financial markets.

Studies of the impact of HFT suggest that information asymmetry between buyers and sellers is reduced over the time and, even if very often HFT are accused of arbitrage, many empirical studies demonstrate a general improvement in market liquidity (as measured by reduction of spreads or increases in depth) and a general reduction of the intraday price volatility. These results ultimately suggest that any regulatory action introduced to curtail this activity may have serious negative implications for liquidity and market participants (Frino et al. 2017), as demonstrated recently both in France and in Italy (Friederich and Payne 2015).

The large volume of papers published on the topic of AT and HFT indicate a clear academic interest in the potential contributions and limitations of HFT activity. However, many open questions remain unanswered:

- It is not clear if the systemic risk is embedded in electronic trading or really caused by HFT.
- The HFT strategies are still partially undiscovered given that the majority of studies use proxies to detect their activity rather than identifiers. Consequently, the latter information might provide a clear evidence of the real impact of the different trading strategies on market quality.
- Only a few studies highlight the effects of HFT activity around specific events like earnings announcements (Frino et al. 2017), news (Scholtus et al. 2014) or macro-news announcements (Bernile et al. 2016) but how do HFTs react around other specific events like mergers and acquisitions or social media releases? What is the effect of their activity on market quality in such cases?
- How do HFTs react to narrative accounting disclosure? Given that the corporate disclosure is moving towards "machine readable" reports, how can firms anticipate HFT trading strategies at the time of disclosure?

Firms and investors should take into consideration that with the advent of HFTs the speed of dissemination of information (earnings, good or bad news, buy or sell quotes and trades) has changed, capital markets have evolved, and complex algorithms may soon become obsolescent.

In 2009 and 2010, HFT techniques were considered a goldmine for sophisticated investors that know how to deploy them against human competitors, however 'dumb' competitors got wise and began to employ similar strategies to defend their wealth stores (*Financial Times* 2017). According to Tabb Group, the US market makers reported $1.1 billion in revenue in 2016, compared with $7.2 billion in 2009. This phenomenon demonstrates a slowdown in the world of HFT with lower profitability. As HFT speed no longer accrues a significant competitive advantage, sophisticated investors are now trying to capture a competitive advantage in predicting markets through quantitative models and artificial intelligence (AI) throwing up new challenges and opportunities for investors, policymakers and scholars alike.

REFERENCES

Aitken, M., Cumming, D., & Zhan, F. (2014). Trade size, high-frequency trading, and colocation around the world. *The European Journal of Finance*, 1–21. https://www.tandfonline.com/doi/abs/10.1080/1351847X.2014.917119.

Aït-Sahalia, Y., & Saglam, M. (2013). *High frequency traders: Taking advantage of speed* (No. w19531). National Bureau of Economic Research.

Aldridge, I., & Krawciw, S. (2017). *Real-time risk: What investors should know about FinTech, high-frequency trading, and flash crashes.* Hoboken, NJ: Wiley.

Allee, K. D., DeAngelis, M. D., & Moon, J. R., Jr. (2018). Disclosure "Scriptability". *Journal of Accounting Research, 56*(2), 363–430.

Angel, J. J., Harris, L. E., & Spatt, C. S. (2015). Equity trading in the 21st century: An update. *The Quarterly Journal of Finance, 5*(1). https://www.worldscientific.com/doi/abs/10.1142/S2010139215500020.

Baron, M., Brogaard, J., Hagströmer, B., & Kirilenko, A. A. (2017). Risk and return in high-frequency trading. *Journal of Financial and Quantitative Analysis (JFQA)*, Forthcoming. Available at SSRN: https://ssrn.com/abstract=2433118.

Bernile, G., Hu, J., & Tang, Y. (2016). Can information be locked up? Informed trading ahead of macro-news announcements. *Journal of Financial Economics, 121*(3), 496–520.

Bessembinder, H., Carrion, A., Tuttle, L., & Venkataraman, K. (2016). Liquidity, resiliency and market quality around predictable trades: Theory and evidence. *Journal of Financial Economics, 121*(1), 142–166.

Bloomfield, R. J. (2002). The "incomplete revelation hypothesis" and financial reporting. *Accounting Horizons, 16*(3), 233–243.

Boehmer, E., Li, D., & Saar, G. (2015). *Correlated high-frequency trading.* Manuscript, Cornell University, Ithaca, NY.

Boehmer, E., Fong, K. Y. L., & Wu, J. (2018). *Algorithmic trading and market quality: International evidence* (AFA 2013 San Diego Meetings Paper). Available at SSRN: https://ssrn.com/abstract=2022034.

Bongaerts, D., Kong, L., & Van Achter, M. (2016). *Trading speed competition: Can the arms race go too far?* Available at SSRN: https://ssrn.com/abstract=2779904.

Bouveret, A., Guillaumie, C., Roqueiro, C. A., Winkler, C., & Nauhaus, S. (2014). High-frequency trading activity in EU equity markets. *ESMA Report on Trends, Risks and Vulnerabilities, 1*, 41–47.

Brogaard, J., Hendershott, T., Hunt, S., & Ysusi, C. (2014). High-frequency trading and the execution costs of institutional investors. *Financial Review, 49*(2), 345–369.

Brogaard, J., Hendershott, T., & Riordan, R. (2017). High frequency trading and the 2008 short-sale ban. *Journal of Financial Economics, 124*(1), 22–42.

Brogaard, J., Carrion, A., Moyaert, T., Riordan, R., Shkilko, A., & Sokolov, K. (2018). High frequency trading and extreme price movements. *Journal of Financial Economics, 128*(2), 253–265.

Brunnermeier, M., & Pedersen, L. H. (2005). Predatory trading. *The Journal of Finance, 60*(4), 1825–1863.

Caivano, V., Ciccarelli, S., Stefano, G. D., Fratini, M., Gasparri, G., Giliberti, M., et al. (2012, December). *High frequency trading: Definition, effects, policy issues.* CONSOB (Commissione Nazionale per le Societá e la Borsa) (Discussion Paper No. 5). pp. 1–60.

Capelle-Blancard, G., & Havrylchyk, O. (2016). *The impact of the french securities transaction tax on market liquidity and volatility.* Available on SSRN: https://papers.ssrn.com/sol3/papers.cfm?abstract_id=2378347.

Carrion, A. (2013). Very fast money: High-frequency trading on the NASDAQ. *Journal of Financial Markets, 16*(4), 680–711.

Chaboud, A. P., Chiquoine, B., Hjalmarsson, E., & Vega, C. (2014). Rise of the machines: Algorithmic trading in the foreign exchange market. *The Journal of Finance, 69*(5), 2045–2084.

Chung, K. H., & Lee, Albert J. (2016). High-frequency trading: Review of the literature and regulatory initiatives around the world. *Asia-Pacific Journal of Financial Studies, 45*(1), 7–33.

Clarke, V., & Braun, V. (2013). Teaching thematic analysis: Overcoming challenges and developing strategies for effective learning. *The Psychologist, 26*(2), 120–123.

Clark-Joseph, A. (2013). *Exploratory trading.* Unpublished job market paper. Harvard University, Cambridge, MA.

Colliard, J. E., & Hoffmann, P. (2017). Financial transaction taxes, market composition, and liquidity. *The Journal of Finance, 72*(6), 2685–2716.

Conrad, J., Wahal, S., & Xiang, J. (2015). High-frequency quoting, trading, and the efficiency of prices. *Journal of Financial Economics, 116*(2), 271–291.

Dumay, J. (2014). 15 years of the journal of intellectual capital and counting: A manifesto for transformational IC research. *Journal of Intellectual Capital, 15*(1), 2–37.

European Union, MIFID II - Directive 2014/65/EU of the European Parliament and of the Council of 15 May 2014 on markets in financial instruments and amending the Insurance Mediation Directive and AIFMD. *Article, 4*(1)(39).

Friederich, S., & Payne, R. (2015). Order-to-trade ratios and market quality. *Journal of Banking & Finance, 50,* 214–223.

Frino, A., Mollica, V., & Webb, R. I. (2014). The impact of co-location of securities exchanges' and traders' computer servers on market liquidity. *Journal of Futures Markets, 34*(1), 20–33.

Frino, A., Mollica, V., Monaco, E., & Palumbo, R. (2017). The effect of algorithmic trading on market liquidity: Evidence around earnings announcements on Borsa Italiana. *Pacific-Basin Finance Journal, 45,* 82–90.

Garfield, E. (1977). Introducing citation classics-human side of scientific reports. *Current Comments, 1,* 5–7.

Goettler, R. L., Parlour, C. A., & Rajan, U. (2009). Informed traders and limit order markets. *Journal of Financial Economics, 93*(1), 67–87.

Goldstein, M. A., Kumar, P., & Graves, F. C. (2014). Computerized and high-frequency trading. *Financial Review, 49*(2), 177–202.

Grant, J., & Rachel S. (2012). *Italy to limit high-frequency orders Financial Times.* Available at: https://www.ft.com/content/1bbcc370-5bb5-11e1-a447-00144feabdc0. Last accessed 16 August 2018.

Haferkorn, M., & Zimmermann, K. (2014). *The German high-frequency trading act: Implications for market quality.* Available at SSRN: https://ssrn.com/abstract=2514334.

Hagströmer, B., & Nordén, L. (2013). The diversity of high-frequency traders. *Journal of Financial Markets, 16*(4), 741–770.

Hasbrouck, J., & Saar, G. (2013). Low-latency trading. *Journal of Financial Markets, 16*(4), 646–679.

Hendershott, T., Jones, C. M., & Menkveld, A. J. (2011). Does algorithmic trading improve liquidity? *The Journal of Finance, 66*(1), 1–33.

Hoffmann, P. (2014). A dynamic limit order market with fast and slow traders. *Journal of Financial Economics, 113,* 156–169.

Jarnecic, E., & Snape, M. (2014). The provision of liquidity by high-frequency participants. *Financial Review, 49*(2), 371–394.

Jones, C. M. (2013). *What do we know about high-frequency trading?* (Columbia Business School Research Paper No. 13-11). Available at SSRN: https://ssrn.com/abstract=2236201.

Kaminska, I. (2017, March 28). HFT as an insight into where fintech is going, *Financial Times.* Available at FT: https://ftalphaville.ft.com/2017/03/28/2186482/hft-as-an-insight-into-where-fintech-is-going/.

Kirilenko, A., Kyle, A., Mehrdad, S., & Tugkan, T. (2017). The flash crash: The impact of high frequency trading on an electronic market. *The Journal of Finance, 72*(3), 967–998.

Malinova, K., Park, A., & Riordan, R. (2013). *Do retail traders suffer from high frequency traders.* Available at SSRN: 2183806.

Massaro, M., Dumay, J., & Guthrie, J. (2016). On the shoulders of giants: Undertaking a structured literature review in accounting. *Accounting, Auditing & Accountability Journal, 29*(5), 767–801.

Menkveld, A. J. (2013). High frequency trading and the new market makers. *Journal of Financial Markets, 16*(4), 712–740.

Menkveld, A. J. (2016). The economics of high-frequency trading: Taking stock. *Annual Review of Financial Economics, 8,* 1–24.

Menkveld, A. J. & Yueshen, B. Z. (2017). The flash crash: A cautionary tale about highly fragmented markets. *Management Science,* Forthcoming. Available at SSRN: https://ssrn.com/abstract=2243520.

Ministry of Economy and Finance (MEF). (2013). *Explanatory memorandum.* Available at: http://www.mef.gov.it/primo-piano/documenti/Relazione_Illustrativa_English_version_6_2_2013.pdf. Last accessed 16 August 2018.

Prix, J., Loistl, O., & Huetl, M. (2007). Algorithmic trading patterns in Xetra orders. *The European Journal of Finance, 13*(8), 717–739.

Rogers, J. L., Skinner, D. J., & Zechman, S. L. (2017). Run EDGAR run: SEC dissemination in a high-frequency world. *Journal of Accounting Research, 55*(2), 459–505.

Scholtus, M., van Dijk, D., & Frijns, B. (2014). Speed, algorithmic trading, and market quality around macroeconomic news announcements. *Journal of Banking & Finance, 38,* 89–105.

Securities, ASIC—Australian, and Investments Commission. (2010). *Report 215: Australian equity market structure.* Available at: https://download.asic.gov.au/media/1343084/rep-215.pdf. Last accessed 16 August 2018.

U.S. Commodity Futures Trading Commission and the U.S. Securities and Exchange Commission (CFTC and SEC). (2010). *Findings regarding the market events of May 6, 2010.* Available at: https://www.sec.gov/news/studies/2010/marketevents-report.pdf. Last accessed 16 August 2018.

U.S. Securities and Exchange Commission (SEC). (2010). *Concept Release on Equity Market Structure 34-61358.* Available at: https://www.sec.gov/rules/concept/2010/34-61358fr.pdf. Last accessed 18 August 2018.

Yao, C., & Ye, M. (2018). Why trading speed matters: A tale of queue rationing under price controls. *The Review of Financial Studies, 31*(6), 2157–2183.

CHAPTER 5

InsurTech

Dominic Cortis, Jeremy Debattista,
Johann Debono and Mark Farrell

Abstract Traditional challenges insurers face include: (1) asymmetric information—the inability to price a policyholder correctly; (2) moral hazards—the change of attitude following cover; and (3) claims management. In this chapter, we discuss how disruptive technologies are evolving in the insurance sector and the challenges faced in their implementation. We show how large and continuous datasets are transforming the general insurance markets and their business processes, as well

D. Cortis (✉)
University of Malta, Msida, Malta
e-mail: dominic.cortis@um.edu.mt

J. Debattista
ADAPT Centre, Trinity College Dublin, Dublin, Ireland
e-mail: debattij@tcd.ie

J. Debono
Birmingham City University, Birmingham, UK
e-mail: johann.debono@mail.bcu.ac.uk

M. Farrell
Queen's University Belfast, Belfast, Northern Ireland, UK
e-mail: mark.farrell@qub.ac.uk

T. Lynn et al. (eds.), *Disrupting Finance*, Palgrave Studies
in Digital Business & Enabling Technologies,
https://doi.org/10.1007/978-3-030-02330-0_5

as enticing desirable policyholder behaviour and streamlining claims management. We then discuss how artificial intelligence is improving traditional insurance processes from the first point of contact to claims management. We also examine the use of artificial intelligence as a means of interacting with prospective clients and existing policyholders. Finally, we explain how blockchain technology can transform the structure of the insurance market to a peer-to-peer format.

Keywords InsurTech · Insurance · Telematics · P2P insurance

5.1   Introduction

It appears that the age-old business of insurance is finally in the throes of change. Incumbent insurance companies are under threat not only from tech giants such as Amazon entering the market (Seekings 2017), but also from agile start-up entities, that are leveraging the power of technology to innovate their way to market share. This utilisation of technology to improve efficiency and savings in underwriting, risk pooling and claims management from the current insurance model has come to be known as "InsurTech", deriving inspiration from the more well-established concept of "FinTech".

This chapter starts off by describing the process and challenges of a traditional insurer. This is followed by a discussion of future developments in InsurTech. These developments are then discussed in light of the big data paradigm, artificial intelligence (AI) techniques and distributed ledger infrastructures (also known as blockchains or distributed ledge technologies [DLT]).

5.2   How Does Insurance Work?

The business of insurance involves risk transfer from the policyholder to the insurer. The insurer pools similar risks in homogenous groups and pays out any claims from the collected premiums and sometimes from own reserves. The consequence of such risk pooling is a lower variability of outcomes and less likelihood of extreme payouts. For example, consider the probability of a delayed flight as 10% and the cost incurred being €100 if this happens. If ten individuals on separate independent flights pool their risk, the probability of all of them having a delayed flight and hence paying out €100 each is 0.00000001%.[1] In a typical scenario, the insurer would charge above €10 (the fair price) as a premium to cater for claims, contingency, management costs and profits.

The insurer goes through the process of underwriting on being approached to provide cover, that is assessing whether the risk should be taken on board and at what price and then in the event a claim is notified, a claims management process is initiated. Two main challenges faced by insurers are 'adverse selection' and 'moral hazard'. The latter refers to the policyholder changing attitude following attainment of insurance cover (e.g., not locking the doors of a property as knowing the property is insured). 'Adverse selection' is typically the outcome of asymmetrical information whereby a policyholder ends up being pooled (and priced) in a specific risk group despite having a riskier profile. This may have been the case whereby the insurer does not use particular information in pricing risk or this information was possibly withheld. Throughout this chapter we discuss how technology is enabling insurers to price policyholders actuarially fair and reduce moral hazard by controls or gentle prompting.

5.3   The Big Data Paradigm

In today's world, data is becoming an indispensable commodity, leading industries to transform their business processes and value chains into data-driven ones. In the insurance industry, one can relate to this "phenomenon" by observing how *historic insurance models* became more adaptive by making provisions for this ever-growing flow of data through various heterogeneous (unstructured/semi-structured) sources such as sensors or social media. This is often called *Big Data*, which is characterised by the 5 dimensions (5Vs): Volume (how much), Velocity (how fast), Variety (different kinds), Veracity (truthfulness and trustworthiness) and Value (the worth of data).

This proliferation of data, or "big data", has allowed the InsurTech businesses and more forward-thinking established insurance companies to harness a unique selling proposition and competitive advantage over

---

[1]On the other side of the scale, the probability of no one having a delayed flight is 90.438% which is lower than 99%.

other market participants. We now examine three different areas in which big data has impacted the insurance world, to date: Telematics, Wearable Technology and the Internet of Things (IoT).

5.3.1   Telematics

A key consideration of insurance is to ensure that it charges adequate premiums by pricing its products appropriately. Within the automotive insurance industry, insurers have, for a long time, proxied the risk of policyholders making an accident related claim via rating factors such as drivers' age, gender,[2] postcode, car model and claims experience. The underlying assumption is that these rating factors are predictive of the likelihood of claim. For example, a young driver with a sports car is deemed to be more likely to be involved in an accident than a middle-aged driver in a sedan and thus is priced accordingly. This pricing mechanism is problematic in the sense that some of the young drivers in question may actually be a much lower risk, in terms of driving ability, irrespective of their car type and age status. This mispricing can lead to adverse-selection where the low-risk individuals move out of the insured pool and seek cover elsewhere, eventually leading to what is known as the adverse-selection spiral. Telematics seeks to overcome this issue by using on-board technology to monitor and assess the driving behaviour of each individual driver, thus moving insurance from a pooled pricing model to a more individual specific model where the underlying risk is more closely monitored (Barbara et al. 2017).

The telematics technology devices (also known as a "black box") can pick up diverse driving metrics such as location, time of day, mileage, driving frequency, behaviour around hazardous zones, speed, rates of acceleration and braking habits. These metrics can then be considered in a more accurate and individualised pricing model, which ultimately allows the previously trapped pooled policyholders to break free from their features such as age and prove their worth as safe drivers that are a good risk and unlikely to have an accident and hence claim. Not all policyholders are bound to profit from this as pricing accuracy may lead in certain individuals being priced out of the market, as the previously good risks that were subsidising their premium, are now priced on a more personalised basis.

---

[2] Although it is now illegal to base any insurance pricing on gender with the European Union.

5.3.2   Wearables

Wearables are typically viewed as being those of a wrist-borne nature (e.g., FitBit and the Apple watch), however, the technology is now generating masses of data from a variety of sensors embedded in devices (e.g., medical technology) to fashion items such as jewellery, clothing and shoes. These devices are becoming more affordable to the general public, and similar to the uptake of telematics, insurers are benefiting from this surge of available data to improve upon their pricing models.

Potential wearable derived biometric information includes those from physical activity (e.g., number of steps, time spent sitting, miles cycled), cardiovascular measures (heart rate, heart rate variability, ECG, blood pressure), sleep data (quantity and quality), body temperature, galvanic skin response, blood sugar and even pollution exposure. It should come as no surprise, therefore, that the main insurance interest from wearable data comes from health long-term care insurers and to a lesser degree life insurance companies.

In a similar fashion to telematics, wearables provide the insurer with a means to determine, or at least get closer to, the true underlying risks of the insured policyholder. The opportunities for wearables, in the insurance world, potentially go beyond that of refinement of current morbidity and mortality models. They also provide the insurer with a means by which they can improve their marketing efforts (e.g., providing free wearables as has already been done in the UK [Stables 2016]), reduce customer churn through greater engagement and touch points (e.g., from monthly premium discounts) and potentially even motivate healthy behaviour change as well as alerting customers to health concerns (and hence reducing moral hazard).

Whilst the opportunities for insurance companies in this space do indeed appear to be great, the nascent nature of wearables for insurance purposes means that there are many issues and considerations to overcome before their use becomes mainstream. Chief amongst these issues is the accuracy of pricing models and reliability of the devices.

5.3.3   Smart Homes and the Internet of Things (IoT)

The network of physical devices embedded with sensors and connectivity, allowing the transmission and communication of data has come to be known as IoT. Applications range from smart home devices (e.g., smoke alarms, thermostats and fridges) to environmental monitoring (e.g., examining air and water quality) and has permeated the market to such an extent that forecasts suggest that by 2020 IoT will consist of as many as 30 billion objects (Nordrum 2016) and will have a global market value of $7.1 trillion (Hsu and Lin 2016).

As per telematics and wearables, IoT also facilitates the provision of a multitude of new data sources of interest to the insurance world. And again, as per telematics and wearables, the opportunities to use this data extend beyond pricing power (e.g., discounts to customers that lock their sensor-based windows and doors when away from the house). Smart home devices, for example, allow the insurer to potentially move towards being proactive in terms of managing risks. The traditional insurance model has been one of zero intervention prior to a claims assessment, but the data from IoT smart home sensors opens the potential for a new type of customer interaction. A relationship whereby the insurance company now takes an active role in engaging with the customer between the point of sale and claim. For example, the data from a sensor monitoring water pressure could be used to alert the policyholder of a leakage problem before substantial damage occurs.

5.3.4   *Big Data: Trustworthiness and Privacy Concerns*

In Big Data, veracity is one of the main Big Data dimensions. Whilst in the past this was frequently overlooked as long as data was being harvested from multiple heterogeneous sources, veracity is nowadays a more pressing issue (e.g., due to an increase in public discourse on fake news). Simply put, veracity deals with data uncertainty due to inconsistencies and deliberate deception. These problems create obfuscated data, hindering accurate and correct future analysis and understanding of data and leading to potential insurance fraud.

A second issue in harvesting data from multiple sources is privacy. Personal data might have been generated (e.g., GPS location from mobile device) or harvested (e.g., social networks) from multiple heterogeneous sources. Ethically, data owners (including insurers) should ask the consent of their customers prior to use this data for analysis purposes. Furthermore, businesses with customers in Europe have to comply with the General Data Protection Regulation (GDPR) which looks after the privacy and protection of an individual, addressing problems such as how personal data can be used in this data-driven technological age

(EU 2016). It will be interesting to follow how insurers may use social credit schemes, which monitor also social circles to create a 'credit score' (Gapper 2018), in their underwriting processes in light of these issues.

5.4   Artificial Intelligence

Insurers have already started embracing the use of AI techniques to make sense out of the big raw data and obtain useful insights. Techniques such as deep learning, neural networks and natural language processing, amongst others, are helping in improving business operations and as a natural consequence customer's satisfaction.

5.4.1   Machine Learning and AI in the Underwriting Process

It is very likely that all underwriters will be using machine learning and AI as predominant technologies behind their underwriting decisions over time. Workflows of Big Data processing techniques and AI algorithms enable underwriters to process and understand far more data than traditional processes as well as provide more accurate underwriting predictive assessments. With more predictive models, underwriters can apply more adequate premiums and thus enabling underwriters to reduce their loss ratios.

Motor insurance premiums are traditionally charged for a predetermined amount for a period of twelve months. This buffet-style approach, where you pay the same amount irrespective of use, would not apply if priced via a telematics device (Azzopardi and Cortis 2013) as explained in the earlier section. This device enables data between the insured vehicle and the insurer's central management system to be sent instantly. This means that with the help of AI techniques, insurance companies develop a system of adaptive continuous pricing, instead of having a one-off yearly payment.

Traditional life insurance underwriting involves an underwriter asking a specific set of questions to predict life events of the proposer/s. Lapetus Solutions, a US-based InsurTech start-up, have developed an AI system and is currently partnering with life underwriters to provide quotations using facial analytics technology. This system comprises sensory analytics as well as dynamic questioning. To receive a quotation, the client just needs to send a self-portrait photograph (a "selfie") and this technology will use the image provided to examine the individual's physical features and determine the health status, disease susceptibility and longevity (Lapetus Solutions 2017). From the photo provided, the facial analytics technology in this system examines a considerable number of regions on the face in order to provide data to underwriters relating to Body Mass Index (BMI), estimated age and smoking indication. Furthermore, the system scientifically formulates specific questions that vary depending on the responses provided. These will provide more insight and veracity into the individual longevity as opposed to the standard questions normally found in life insurance proposal forms. The advantage of such system is that the whole process only takes a few minutes to complete.

The insurance industry has also started to adopt the use of AI in health and accident insurance. Innovations such as implanted sensors and wearables that make use of AI provide insurers with valuable data regarding the insured's health. This AI technology would also be advising and educating customers about bad lifestyle choices which may ultimately lead to lower costs for both policyholders and insurers.

5.4.2   AI in Claims Management Process

Presently, AI is also taking over administration associated with run of the mill claims. In one particular case in Japan, an AI system has replaced a team of 30 employees calculating payouts for policyholders (McCurry 2017).

This is not all doom and gloom from an employment perspective as developments may lead to claims handlers dealing with the more challenging claims rather than the tedious ones. For example, Lemonade, a start-up property insurance company, has developed an automated claims process and filing a new claim became relatively easy. The smartphone application will ask the policyholder some generic questions to gather basic information on the claim. The insured does not have to complete a claim form but provide a summary of the claim such as what property was damaged, through the smartphone camera. The data provided will be analysed by the AI and run through 18 anti-fraud algorithms. Non-complex claims are approved within seconds whilst complicated claims are handed over to humans in the claims department.

The use of technology is not only pointing out but also easing the management of complex claims such as, making use of drones to take aerial photos of significant property damage and image analytics to quantify the extent of the damage (Cognizant 2017).

Fraud is reported to represent 10% of all claims in Europe (Insurance Europe 2013). Rather than relying on experienced humans to detect fraud, insurance companies are opting for AI to investigate certain dishonest patterns. An Australian tech company developed an AI fraud investigation system that provides support to insurance companies in the detection of fraudulent claims. Whilst the scope of this AI technology is to assist claims handlers in preventing fraud, it also reduces administration costs for insurance companies. The system can investigate social media accounts, criminal records, property and vehicle history and other documentation submitted with the claim, thus enabling more time for the claims department to analyse results and close claims in a timely manner. Similar services that weed out possible fraudulent policyholders at underwriting stage are available in the market (e.g., ThreatMetrix).

5.4.3   AI in Customer Interaction

The insurance industry is seldom considered as the most innovative with respect to customer services. Despite this view, some insurance companies started to make some progress and introduced the use of chatbots in their day-to-day operations in a similar fashion to other industries. Chatbots are an AI system that are normally linked with messaging applications such as Facebook Messenger, with the main purpose of interacting with existing and prospective clients, thus acting as a virtual customer service representative. These chatbots would interact with clients by determining which insurance products would be most appropriate based on their requirements and answer queries using natural language.

Having chatbots that are powered by AI enables digital interaction with policyholders and prospective clients simpler and faster than with a human element. In recent years, insurance intermediaries have also invested in chatbots, for example, to provide real-time insurance quotation comparisons through messaging applications. Such chatbots can also provide clients with recommendations as to what insurance product would best suit their needs.

5.5   Distributed Ledger Technologies

Distributed Ledger Technologies provide opportunities for disruptive developments within the insurance industry. DLTs can be beneficial to current insurers in their processes but could also create a competitive form of peer to peer insurance networks.

5.5.1   Improving Current Processes Using DLTs

DLTs aim to transform the process of verifying not only transactions (like in cryptocurrencies) but also verification of identities and smart contracts.

Mainelli and von Gunten (2014) summarise the effect of DLTs for insurance within four domains: identity, time, space and mutuality. Using DLTs through the identity checking can improve the underwriting and Know Your Customer requirements for an insurer (Mainelli and Smith 2015; Mainelli and von Gunten 2014). These are particularly useful for a product, individual or data that has gone through a chain of custody or changes. Identity improvements could also potentially limit multiple claims for the same incident. For example, in the case of a travel delay, the policyholder would not be able to claim twice for the same incident.

Consider an alternative scenario whereby health and financial transactions are shared over a DLT. At the policyholder's permission, the insurer may be able to quote a premium for health insurance without the need to provide the data from scratch by filling in forms. Moreover, creating models of this personal data and analysing it together with external data (e.g., from personal wearables provided by the insurance company), this enables real-time adjustments to coverage and pricing, hence reducing the time cycles of insurance products. Conversely, DLTs may lead to a lengthening of time as transactions recorded on DLTs cannot be erased or changed, altering the general perception of longer term contracts.

Whilst the current insurance business model is localised as each insurance product is developed by country, market and region; DLTs are distributed over a network of computers. This may increase the space 'covered' by insurance worldwide. For example, Lorenz et al. (2016) argue that DLTs may be particularly useful in microinsurance within emerging markets, citing the example of crop insurance for farmers. Any such coverage may be automated as claims are paid automatically due to weather conditions without the need of an on-the-ground evaluation. The rewards of DLTs within claims management are not restricted to esoteric insurance practices. The implementation of a smart contract would imply an immediate payment on the delivery of parts following a motor vehicle damage claim (Mainelli and von Gunten 2014).

5.5.2   *P2P Insurance*

The diminishment of space may result in the generation of more peer-to-peer (P2P) insurance practices. This would act like a mutual, being Mainelli and von Gunten's (2014) fourth domain of possible effect of a DLTs on insurance. This may lead to a reshaping of the insurance industry for particular coverages in the same manner that AirBnB and Uber/Grab have disrupted their respective industries. Mainelli and von Gunten (2014) explain that P2P networks may lead to some insurance to function like a Protection and Indemnity Club (P&I Club) rather than a mutual. Considering the example of a delayed flight, costing €10 per person each, a group of ten independent individuals may be able to create a mutual coverage. In a P&I Club format, they would pass some process to join, pay a certain amount per year (say €10) and be ready to add supplementary funds should this not be enough. For example, if a total of two claimed, then the claim costs would be €200, meaning that every participant should pay in an additional €10. If no one claimed, a profit of €100 is shared between the participants (€10 each).[3]

We also envisage the possibility of a similar P2P structure through DLTs with reinsurance added to cover in case that the total premiums do not cover the claims. This would replace the need to add funds should the reserves not be enough to pay out claims.

5.6   CONCLUSION

Insurers have been accused of being slow to adapt to new technologies as 50–70% of insurers' IT budget is spent on running costs rather than research and development (Acord and Equinix 2014). It is clear that Big Data is already and has further potential to revolutionise the insurance business. Areas such as telematics, wearables and IoT are providing a plethora of data which, when combined with advances made in AI, are enabling a much more personalised product to be developed for the consumer. Furthermore, a very different customer relationship is starting to emerge, from these data sources, allowing the insurer to engage more meaningfully with policyholders. Consequently, big data is likely to

---

[3]We are ignoring management costs or investments profits for simplification purposes.

continue to transform the industry on a large scale for the foreseeable future. DLTs, the IoT and AI are starting to disrupt the insurance market as we know it today. The extent to which disruption occurs and how exactly the disruption will happen remains to be seen. However, it seems InsurTechs will play an increasing role as the digital innovation in the insurance world unfolds.

References

Acord and Equinix. (2014). Challenge to change Part 2—The impact of technology. Available at: https://www.equinix.ie/resources/whitepapers/challenge-to-change-part-two-impact-of-technology/. Last accessed 17 August 2018.

Azzopardi, M., & Cortis, D. (2013). Implementing automotive telematics for insurance covers of fleets. *Journal of Technology Management & Innovation, 8*(4), 59–67. https://doi.org/10.4067/s0718-27242013000500005.

Barbara, C., Cortis, D., Perotti, R., Sammut, C., & Vella, A. (2017). The european insurance industry: A PEST analysis. *International Journal of Financial Studies, 5*(2), 14. https://www.mdpi.com/2227-7072/5/2/14.

Cognizant. (2017). Cognizant and measure partner to deliver a dramatic business impact for insurers using drones. Available at: https://www.cognizant.com/industries-resources/insurance/drones-solution-overview.pdf. Last accessed 29 April 2018.

European Union. (2016). Council directive (EU) 5419/2016 on general data protection regulation. Available at: https://eur-lex.europa.eu/legal-content/EN/TXT/?uri=consil%3AST_5419_2016_INIT. Last accessed 17 August 2018.

Gapper, J. (2018). Alibaba's social credit rating is a risky game. *The Financial Times.* Available at: https://www.ft.com/content/99165d7a-1646-11e8-9376-4a6390addb44. Last accessed 27 April 2018.

Hsu, C. L., & Lin, J. C. C. (2016). An empirical examination of consumer adoption of internet of things services: Network externalities and concern for information privacy perspectives. *Computers in Human Behavior, 62,* 516–527.

Insurance Europe. (2013). The impact of insurance Fraud. Available at: https://www.insuranceeurope.eu/sites/default/files/attachments/The%20impact%20of%20insurance%20fraud.pdf. Last accessed 27 April 2018.

Lapetus Solutions. (2017). Lapetus solutions launches new data collection initiative to support the global expansion of its predictive technologies. Available at: https://www.lapetussolutions.com/press/lsipr002/. Last accessed 25 April 2018.

Lorenz, J.-T., Münstermann, B., Higginson, M., Olesen, P. B., Bohlken, N., & Ricciardi, V. (2016). Blockchain in insurance—Opportunity or threat? Available at: https://www.mckinsey.com/industries/financial-services/our-insights/blockchain-in-insurance-opportunity-or-threat. Last accessed 18 March 2018.

Mainelli, M., & Smith, M. (2015). Sharing ledgers for sharing economies: An exploration of mutual distributed ledgers (Aka Blockchain Technology). *Journal of Financial Perspectives, 3*(3). Available at SSRN: https://ssrn.com/abstract=3083963.

Mainelli, M., & von Gunten, C. (2014). Chain of a lifetime: How blockchain technology might transform personal insurance. http://archive.longfinance.net/long-finance-report/903-chain-of-a-lifetime-how-blockchain-technology-mighttransform-personal-insurance.html.

McCurry, J. (2017, January 5). Japanese company replaces office workers with artificial intelligence. *The Guardian.* https://www.theguardian.com/technology/2017/jan/05/japanese-company-replaces-office-workers-artificialintelligence-ai-fukoku-mutual-life-insurance. Last accessed 12 October 2018.

Nordrum, A. (2016). Popular internet of things forecast of 50 billion devices by 2020 is outdated. *IEEE Spectrum, 18.*

Seekings, C. (2017). Amazon set to shake-up UK insurance. *The Actuary.* Available at: http://www.theactuary.com/news/2017/11/amazon-set-to-disrupt-uk-insurance/. Last accessed 17 August 2018.

Stables, J. (2016). Free coffee and cheap Apple Watches: Vitality extends wearable tech discounts. *Weareable.com.* Available at: https://www.wareable.com/apple/vitality-extends-its-wearable-tech-offering-3297. Last accessed 17 April 2019.

CHAPTER 6

Understanding RegTech for Digital Regulatory Compliance

Tom Butler and Leona O'Brien

Abstract This chapter explores the promise and potential of Regulatory Technologies (RegTech), a new and vital dimension to FinTech. It draws on the findings and outcomes of a five-year research programme to highlight the role that RegTech can play in making regulatory compliance more efficient and effective. The chapter presents research on the Bank of England/Financial Conduct Authority (FCA) RegTech Sprint initiative, whose objective was to demonstrate how straight-through processing of regulations and regulatory compliance reporting using semantically enabled applications can be made possible by RegTech. The chapter notes that the full benefits of RegTech will only materialise if the pitfalls of a fragmented Tower of Babel approach are avoided. Semantic standards, we argue, are the key to all this.

Keywords RegTech · FinTech · Semantic technologies · Standards

T. Butler (✉) · L. O'Brien
BIS, University College Cork, Cork, Ireland
e-mail: TButler@ucc.ie

© The Author(s) 2019
T. Lynn et al. (eds.), *Disrupting Finance*, Palgrave Studies in Digital Business & Enabling Technologies,
https://doi.org/10.1007/978-3-030-02330-0_6

6.1   Introduction

RegTech is information technology (IT) that (a) helps firms manage regulatory requirements and compliance imperatives by identifying the impacts of regulatory provisions on business models, products and services, functional activities, policies, operational procedures and controls; (b) enables compliant business systems and data; (c) helps control and manage regulatory, financial and non-financial risks; and (d) performs regulatory compliance reporting. In reference to the previous generation of RegTech (and FinTech) Law Professor Kenneth Bamberger points out that:

> While these technology systems offer powerful compliance tools, they also pose real perils. They permit computer programmers to interpret legal requirements; they mask the uncertainty of the very hazards with which policy makers are concerned; they skew decision-making through an "automation bias" that privileges personal self-interest over sound judgment; and their lack of transparency thwarts oversight and accountability. These phenomena played a critical role in the recent financial crisis. (Bamberger 2009, pp. 669–670)

As Bamberger notes, in the rush to embrace new technologies, the downside risks are either ignored, played down, or transferred. This is apparently so with RegTech, as Patrick Armstrong, Senior Officer for Financial Innovation at the European Securities and Markets Authority (ESMA), warned "it is not without risks" and financial institutions must take care to "delegate tasks, not responsibility...for compliance and risk management activities. Instead the ultimate responsibility remains with the regulated financial institution" (McNulty 2017). There is a dearth of IS research in this new important area of study; however, research in other disciplines is also deficient. Few academic researchers have adopted a critical stance (Packin 2018), while others fail to question the risks associated with this new paradigm (Arner et al. 2016), including the real possibility of creating a digital *Tower of Babel* (Butler 2017). This is, perhaps, the major issue confronting the financial industry.

The industry has, according to Andrew Haldane, created a *Tower of Babel*, which refers to the absence of a "common language" in the financial industry, and the existence of heterogeneous terms and concepts to describe similar business objects, processes, and products (Haldane 2012). This problem permeates the industry down to individual financial institutions, where products, concepts, and terms have different meanings in and across business functions and communities of practice. The emergence of FinTech and RegTech will do little to solve fundamental problems if the industry ends up with a digital *Tower of Babel* that simply digitises the status quo. The essential issue of arriving at shared business and regulatory terminological dictionaries, thesauri, and taxonomies is a huge challenge and a significant obstacle for RegTech. Only then will the issue of *semantic interoperability* be addressed—that is, the capacity of information systems to exchange data with unambiguous, shared meaning. "Semantic interoperability ensures that these exchanges make sense—that the requester and the provider have a common understanding of the "meanings" of the requested services and data" (Heiler 1995, p. 271). However, even if a "common language" did exist in a financial institution, we are still left with what Bamberger (2009) termed the problem of "translation"—that is the gap between the meanings accorded to business concepts by business professionals and the intended behaviours of automated processes, and the meanings recorded by systems analysts and software engineers in data stores and the behaviours embedded in software code. Bamberger demonstrated how the "translation problem" resulted in financial, risk and compliance applications and systems that masked risk, led users into a false sense of security, and provided erroneous signals to business managers resulting in poor decision-making on key issues.

The remainder of this paper is structured as follows. First, we discuss the business drivers for the adoption of RegTech, which incorporates a regulatory perspective. Then we present our empirical case study. We finish by discussing the implications of our research and offer concluding comments.

6.2   Business Drivers of RegTech

The costs of regulatory compliance for the financial industry represent the primary drivers for RegTech adoption. Research published by *The Trade* indicates that banks spent over $100 billion on regulatory compliance in 2016 alone and this cost is rising (McDowell 2017). Bain & Co. estimates that governance risk and compliance (GRC) expenditure accounts for 15–20% of "run the bank cost" and 40% of "change the bank costs" (Memminger et al. 2016). Looking at specific regulations, Dodd Frank has cost $36 billion to date, while MiFID II has cost a €2.5 billion

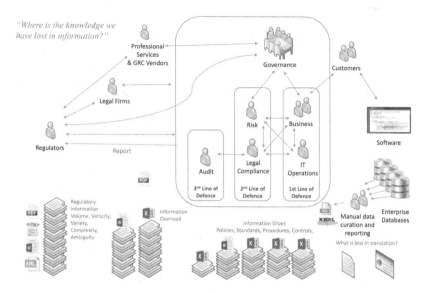

Fig. 6.1 Information overload, complexity, silos, and loss

to date—the latter cost is sure to rise significantly. Thus, given the existing trend, it is estimated that the cost of regulatory compliance will rise from 4 to 10% of revenue in financial institutions by 2022.

At a practical level, it is the volume of regulations that is driving costs. Take, for example, that over 50,000 regulations were published between 2009 and 2012 in the G20. Furthermore, over 50,000 regulatory updates were published in 2015 alone, 100% up on 2012. The scale of the paper mountain for firms is breath-taking: The FCA Handbook stands over 2 metres high; the US Dodd Frank Act has generated over 22,000 pages of provisions; the EU's MiFID II has approximately 30,000 pages of related texts in approximately 1.5 metres paragraphs. Each week sees an average of 45 new regulatory-related documents issued (JWG 2017). Thus, the volume, variety, velocity, and complexity of regulation is challenging firms. Figure 6.1 captures graphically the current approach to regulatory compliance and reporting. Information overload, multiple communication paths, information silos, and manual data curation all increase the risk of information loss and reduced empirical fidelity while driving significant cost.

Another driver is the relative complexity of financial institutions in terms of their business models, legal entity structures, processes, products, services, and markets served. Monitoring, interpreting, and complying with the current and planned regulations is a challenge, even for the largest banks. For smaller firms the costs and complexity may become prohibitive (Walker 2018). The move from people-based solutions to RegTech is reflected in the fact that the aforementioned costs are being expended on consultants, professional services, and IT vendors (Marenzi 2017).

The final drivers are information and data related. Unpublished research from a 5-year-long empirical study by the authors on the issue, involving UK and US regulators and financial institutions, indicates that financial institutions are challenged by their inability to understand:

1. Regulatory requirements and compliance imperatives;
2. The impacts of such regulations on functional activities, policies, and procedures;
3. The changes that are required to business processes and activities;
4. The risks associated with financial products and related business models;
5. The implications for IT systems in terms of data governance and analytics; and
6. How to meet consumer needs, while protecting their rights.

It was in this context that RegTech was first identified as a separate, but emerging industry sector in the financial industry, in the UK Treasury's 2015 Budget Report and subsequently explored comprehensively in the UK's Government Organisation for Science (Walport 2015). UK regulators took note of this new perspective on technology-based solutions for the myriad of problems facing the financial industry and have responded accordingly. In 2016, for example, and in the context of the FCA's Innovation Hub and its Project Innovate, the FCA identified a number of candidate FinTech and RegTech solutions and how they might be used.

Given the significant challenges facing both regulators and the regulated, Christopher Woolard, Director of Strategy and Competition at the FCA, identified several use cases for and capabilities of RegTech (Woolard 2016) viz.

1. First, making the business of complying with reporting requirements simpler.
2. Second, technology that drives efficiencies in regulatory compliance by seeking to close the gap between the intention of regulatory requirements and the subsequent interpretation and
   implementation within firms.
3. Third, technology that simplifies and assists firms in managing and
   exploiting their existing data, supporting better decision-making
   and finding those who are not playing by the rules easier.
4. Finally, technologies and innovations that allow regulation
   and compliance processes to be delivered differently and more
   efficiently.

The above conceptualisation indicates a role for several new technologies
including artificial intelligence (AI), blockchain/distributed ledger technologies (DLT), machine learning, natural language processing (NLP),
and data analytics. However, it was clear from several presentations at the
European Central Bank Data Standards for Granular Data Conference
2017 that the *sine qua non* for the success of RegTech would be the use
of open standards and semantic technologies (Palmer 2017). A standards-based approach would be necessary to address the core issues of
the translation and *Tower of Babel* problems (Bamberger 2009; Haldane
2012; Butler 2017).

6.3   REGTECH IN FOCUS: DIGITAL REGULATORY REPORTING

The need for an industry-wide standards-based approach to regulatory
compliance and reporting, articulated in the position paper of Butler
(2017) on open standards for RegTech, found purchase with both UK
regulators and the financial industry. The Bank of England and the FCA
subsequently hypothesised that standards-based RegTech could help
automate, and make more efficient and cost-effective, the task of regulatory reporting by financial institutions. To confirm their hypothesis, the
UK regulators instituted a Technology Sprint—in this case the RegTech
Sprint.

The remainder of this chapter presents a short case study of the
RegTech Sprint. This was undertaken by the Bank of England and the

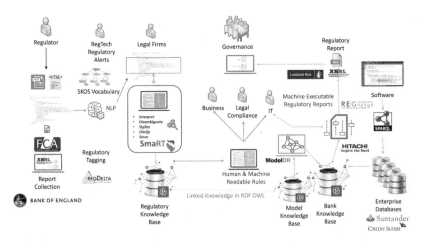

Fig. 6.2 Digital Regulatory Reporting

FCA in conjunction with over 50 participants from across the financial industry in the UK. The Sprint took place during the last two weeks of November 2017. Participants included regulators from the FCA, Bank of England, industry firms Santander, Credit Suisse, Hitachi Vantara, Lombard Risk, Model Drivers, Regnosys, JWG-RegDelta, Governor Software, law firms Linklaters and Burgess Salmon, academics from Yale, and the GRC Technology Centre at University College Cork.

The primary objective of the RegTech Sprint was to provide a Proof of Concept (PoC) that demonstrated the feasibility of the straight-through processing of regulations and semi-automated regulatory reporting—this process is termed Digital Regulatory Reporting. Figure 6.2 presents the key activities that realised this objective.

The first step in Digital Regulatory Reporting is to digitise the regulatory provisions. As indicated in Fig. 6.2, the FCA currently publishes its Handbook of Regulations in HTML and PDF formats. Key concepts are linked using Hypertext. In the case of the FCA, specific FCA concepts are defined in the Handbook Glossary. In its current form, the Handbook provisions and rules are not readily machine-readable or machine-executable. The first part of the PoC was to investigate how AI could be employed to process regulatory provisions and provide Regulatory Alerts.

6.3.1   Phase 1: Digital Regulatory Alerts

RegTech vendor RegDelta has developed taxonomies of regulatory topics using the World Wide Web Consortium (W3C)'s Simple Knowledge Organisation System (SKOS)[1] and AI to semantically tag regulatory provisions to indicate their scope and application so alerts could be generated. SKOS is based on the W3C's Resource Description Framework (RDF),[2] and is one of the three foundational Semantic Web technologies, the other two being SPARQL and the Web Ontology Language (OWL).[3] SKOS enables organisation to transcend the limitations of business glossaries to create taxonomies and thesauri that are both human- and machine-readable. Semantic tagging of regulations is but the first step in the regulatory compliance process and the SKOS namespace helps address the *Tower of Babel* problem. This is an example of the straightforward application of AI and semantic technologies to help manage the volume and complexity of regulations by having a machine answer the *what* and *which* questions. That is, *what* are the themes in regulatory provisions and *which* activities and products do they target. While this process helps partially digitalise regulatory provisions, Sprint participants recognised that the source regulations would have to be redrafted and captured in RDF, if the objective was to be achieved. This task fells to the SmaRT application.

6.3.2   Phase 2: Making Regulations Digital

The core semantic technologies in SmaRT are based on W3C and industry standard semantic technologies. SmaRT applies the Semantics of Business Vocabulary and Business Rules (SBVR) standard proposed by Object Management Group. SBVR enables business subject matter experts to capture and express their vocabularies and rules in a systematic way according to the precepts of first-order deontic-alethic logic. The SmaRT application applies its Mercury implementation of SBVR Structured English to capture the meaning of the tagged regulatory provisions in vocabularies (*alethic logic*, indicating a necessity, possibility, or impossibility) and rules (*deontic logic*, i.e. what should be, specifying

[1] https://www.w3.org/2004/02/skos/.

[2] https://www.w3.org/RDF/.

[3] https://www.w3.org/2003/Talks/0617-Munchen-IH/25.html.

obligations, prohibitions, permissions, etc.). This approach helps formalise regulatory and legal knowledge in the form of vocabularies (a la SKOS) and rules explicitly expressed to identify whether a regulatory provision prohibits certain behaviours, processes or products, or permits them, or obliges firms to behave in a certain way towards customers, and so on. SmaRT regulatory knowledge is persisted in an RDF Knowledge Base to make it machine-readable and machine-computable. AI-based inference and reasoning capabilities are provided by the SmaRT Ontology expressed in OWL.[4] An OWL ontology permits granular semantic querying (using SPARQL) and reasoning to identify new or consolidate existing knowledge—it helps bring semantic interoperability to traditional data systems.

In order to achieve the objective of the Sprint, the Bank of England and the FCA chose to use the Supervision Reporting Requirements provisions in the FCA Handbook (Sup 16.12) along with supplementary definitions supplied by regulators from the Bank of England. Sup 16.12 instructs financial institutions how to report relevant financial data to the UK supervisory authorities—the Bank of England and the Prudential Regulatory Authority (PRA) in this case. The problem with the FCA Handbook, and indeed regulatory rules in general, is that they tend to be ambiguous, drafted in legalese, with links to parent and related legislation, references to financial and technical standards, and links to relevant guidelines.

During the Sprint, the SmaRT application enabled legal and financial industry experts to transform complex legislation in Sup 16.12, related regulatory rules, and other texts containing standards and guidelines, into a human-readable regulatory natural language (RNL) This standards-based RNL is captured by SmaRT using a combination of human knowledge and expertise, augmented by AI and Machine Learning technologies, such as those in RegDelta, and presented to users in an HTML- and XML-based web interface.

Thus, Sprint participants encoded the regulatory provisions into SmaRT's vocabulary and rules in a human-readable format. However, they are also automatically persisted in a machine-readable format in the SmaRT Regulatory Knowledge Base in RDF. As indicated, SmaRT captures the relationships between concepts in RDF triples (e.g. investments

---

[4]https://www.w3.org/OWL/.

Fig. 6.3 SmaRT rules from Sup 16.12

firm manufactures financial products). The SKOS namespace is limited to expressing taxonomic and related categorical relationships. SmaRT's namespace is much richer. In Fig. 6.3, the relationship RAG must submit FSA001 is captured as an RDF triple. Triples are declarative axioms and the building blocks of SmaRT rules. SmaRT rules are represented as RDF/Turtle graph pattern.[5] Thus expressed, data can be checked for consistency or compliance with such rules. Standard W3C technologies such as SPARQL,[6] SPIN,[7] and SHCL[8] are employed for this purpose. SPARQL (SPARQL Protocol and RDF query language), is, as its name indicates, the W3C query language for the Semantic Web and siloed and distributed networked systems. For example, SPARQL can be used to enable querying and integration of siloed financial and risk data for regulatory reporting and risk management. Rules such as those present in Fig. 6.3 are the constituent elements of complex regulatory provisions expressed in a human-readable format in HTML and in a

[5] https://www.w3.org/2001/sw/DataAccess/rq23/#BasicGraphPatternMatching.
[6] https://www.w3.org/TR/rdf-sparql-query/.
[7] https://www.w3.org/Submission/2011/SUBM-spin-overview-20110222/.
[8] https://www.w3.org/TR/shacl/.

machine-readable format as an RDF graph pattern. The closest analogue to this approach is the relationship between Wikipedia and DBpedia.[9] The human-readable Wikipedia content is expressed in HTML and rendered into web page in a browser; however, DBpedia captures Wikipedia content and data in RDF, with concepts modelled in an ontology. Thus, information contained in related/linked Wikipedia pages can be queried (using SPARQL), extracted, federated, integrated, relationships uncovered or inferred, and new knowledge created.

6.3.3   Phase 3: Performing Digital Regulatory Reporting

The knowledge embedded in the SmaRT RDF-based vocabulary was used by software engineers from Hitachi, Regnosys, and Lombard Risk to map firm-specific data concepts in the anonymised customer account data supplied by Santander to equivalent concepts in the Regulatory Knowledge Base. The RDF-based rule graph patterns were employed to create SPARQL queries to extract compliant data on retail customer accounts. Using the SPARQL Inference Notation (SPIN) framework, rules can be graphed and executed. A software application was created to automate this process. This was then used to extract the required data, transform and load it, and then perform the required calculations and populate relevant cells in the FSA 001 Balance Sheet form for submission to the Bank of England.

The major achievement in executing the PoC came when the rule governing customer account reporting was changed. Once the rule change was captured in SmaRT and expressed in RDF, the software application executed over the changed rule (as an RDF graph pattern) and then populated the appropriate fields in FSA 001 form with the required data. No change in the software algorithm was required. This was a major development and provided the PoC.

6.3.4   Phase 4: Creating Meta-Data Models
for Semantic Interoperability

Referring back to Fig. 6.2, Model Drivers (ModelDR) demonstrated a key benefit of the above approach to solving the semantic interoperability problem discussed earlier. ModelDR semantic modelling application

---

[9] https://en.wikipedia.org/wiki/DBpedia.

was able to create ontology-based meta-data models, based on SmaRT vocabularies and rules, that will help scale up the findings and make Digital Regulatory Reporting a reality in the Enterprise. During the Sprint, the ModelDR application was integrated with SmaRT in order to demonstrate how SMEs could capture domain knowledge (here on regulatory provisions) and use this as an input to semantic models expressed in OWL—such models may be used for data virtualisation (see Kontchakov et al. 2013). Such models are currently being built at great cost by major banks. The ability to have business professionals participate in this process is argued to make this process more efficient and help address the aforementioned translation problem and make semantic interoperability possible. This approach also helps scale up the PoC to a working enterprise-wide solution.

6.4   Discussion and Implications

While Know Your Customer (KYC), Anti-money Laundering (AML), and the financial crime problem spaces are, perhaps, the most mature areas in the application of RegTech, Enterprise Data Management is, perhaps, the most important area for the application of RegTech and the lens through which all RegTech solutions should be viewed. The rationale for this assertion is simple and straightforward—financial enterprises have become more or less fully digitised and almost all people, business objects and processes are represented in and through data, be it structured or unstructured. Regulations themselves are unstructured data—although, regulators are seeking to bring structure to regulatory provisions and rules. In the area of conduct risk, for example, predictive analytics and machine learning are being used to identify insider (cyber) threats, suspicious activity (fraud and financial crime), insider trading, and employee misconduct, all based on data captured from phone calls, emails, business transactions, and so on. Unfortunately, the approaches being taken by multiple vendors using proprietary approaches, and applying technologies as diverse as AI, machine learning, NLP, DLT, biometrics, cryptography, cloud computing, and open APIs, may result in a digital *Tower of Babel*, as *semantic interoperability* a major issue for the industry.

The traditional approach employed in Bamberger's *Technologies of Compliance*, whether GRC or RegTech, is to transform and map regulatory provisions, compliance imperatives and rules into software code.

This approach is evident in early RegTech solutions in the AML/KYC/Financial Crime domains. Thus, financial institutions adopting such technologies face a "black box" solution, with an attendant regulatory risk that a client will, for example, be on-boarded in breach of governing regulations. This might happen if a regulatory rule is not properly encoded or if all permutations and combinations are not accounted or tested for. Depending on the gravity of the breach, a financial institution could find itself with a hefty fine or risk of being put out of business. In the course of our research, lawyers critical of RegTech put this argument forward and criticised the often unquestioning acceptance of vendor claims by financial institutions as to the abilities of their software applications to automate or support decision-making around KYC and client on-boarding. RegTech vendors cite Intellectual Property (IP) considerations for their unwillingness to disclose what is in their black boxes (Tyler 2017). What is required here by financial institutions and regulators is provenance between the original provisions/rules and their instantiation in computer algorithms. The RegTech Sprint PoC provides evidence of the utility of a human- and machine-readable intermediate language in ensuring faithful translation between the source regulations, the interpreted provisions, and their representation in the software that underpins technologies of compliance.

Echoing Kenneth Bamberger's argument presented at the beginning of this chapter, Packin (2018, p. 194) warns of the downside to RegTech and argues that its adoption "requires a carefully tailored design of the technology, a joint effort of the regulators and the private sector and some shifts in corporate thinking." Evidence from the RegTech Sprint indicates that this is underway. However, the RegTech genie is out of the bottle and a major problem facing the industry is the growing number of proprietary RegTech solutions from multiple vendors, none of which are aligned around a common model or infrastructure. It has been brought to our attention by executives from Globally Systemic Important Banks (GSIBs or GSIFIs) that the last thing they want is to have multiple FinTech and RegTech solutions, from multiple vendors, adding to the proliferation of applications across their institutions and to the burgeoning "spaghetti pots" of code and data.

The FCA is advocating the adoption of open source technologies and open semantic standards, such as those developed by the W3C, to link and make machine-readable and machine-executable structured and unstructured data across heterogeneous sources. Hence, its focus is on

XML/RDF (and also Turtle and JSON-LD), ontologies (in OWL), and related standards, such as the SBVR, and its extensions, to express regulatory vocabularies and rules in order to underpin the semantic interoperability of systems.

It is also clear from our experience that large financial institutions are beginning to address the core problems of data governance and data virtualisation using semantic technologies that enable interoperability between systems. SBVR is being used by major banks to help map regulatory concepts on to business concepts. Ontologies are being used as meta-data models hosted in RDF triple stores as knowledge bases for data extracted from heterogeneous relational data stores and other sources. This semantic approach to data virtualisation uses SPARQL to field federated queries over the distributed meta-data/data in relational data stores. The operational data stays where it is, with the data of interest returned from multiple data sources, integrated using the ontologies (as meta-data models), with further analysis and processing performed in an RDF triple store. This approach takes on an AI dimension when inferencing engines or reasoners are employed to add knowledge to a knowledge base. A semantic reasoner or rules engine consists of algorithms that infer logical conclusions from a set of asserted axioms or facts expressed in RDF/OWL. From a data perspective, previously unknown or unrecognised relationships across heterogeneous data sets can be asserted, thus adding more knowledge. Successful ontology-based solutions already exist in a wide variety of domains from defence and intelligence, to capital markets, to regulatory compliance (Palantir 2018).

From a business perspective, this approach enables regulatory semantics (vocabularies and rules underpinning regulatory provisions and compliance imperatives) to be mapped to business semantics (vocabularies and rules expressed in business policies, operational standards, controls through to meta-data repositories/data dictionaries). It also permits the disambiguation, extraction, and integration of heterogeneous firm-specific data for regulatory compliance reporting and risk management. Capturing business semantics in a knowledge base is the *sine qua non* of good data governance. An example here is the Model Knowledge Base referred to in Fig. 6.1. Linking a model knowledge base like this with a regulatory knowledge base and integrating both with a business knowledge base can enable straight-through processing of regulations and automated regulatory compliance reporting of both financial data and non-financial data. Interestingly, industry bodies like the IFRS

Foundation/International Accounting Standards Board are using semantic technologies and ontologies to make their standards and reporting XBRL taxonomies both human- and machine-readable. So too are the ISO20022/SWIFT initiatives in their efforts to enhance its financial messaging standard. Thus, a point of convergence is not too far off, and firms across the financial industry need to ensure that they are in a position to capitalise on the very real benefits offered by semantic technologies for FinTech and RegTech.

6.5   Conclusion

This chapter draws on recent field research to demonstrate the promise and potential of RegTech. However, it adds a cautionary note that the full benefits of RegTech will only materialise if the pitfalls of a fragmented *Tower of Babel* approach are avoided (Butler 2017). Semantic standards are the key to all this.

It is clear that the benefits of RegTech go well beyond straight-through processing of compliance reporting of financial data, such as balance sheet reporting and the quantification of organisational or systemic financial risk. We have argued that one of the benefits of the application of semantic standards is data governance, through the ability to create machine-readable meta-models that enable data virtualisation across heterogeneous data stores. This approach may make the enterprise data warehouse a thing of the past. However, it also enables an enhanced data-driven approach to the management of non-financial risk and associated regulatory compliance reporting. Here, data from siloed, heterogeneous databases can be virtualised and ontologies and/or predictive analytics/machine learning algorithms and AI applied to identify insider or cyber threats, suspicious activities, financial fraud by customers/clients, and a wealth of other applications. Readers will begin to realise that in this context RegTech can be applied across industry sectors, and not just the financial industry, as non-financial risks, such as operational risk and employee misconduct, consumer protection are not industry specific.

The financial industry spends more on IT and data than any other, over $360 billion per annum according to Gartner. Given the fundamental problems, it faces with regulatory compliance, which is costing the industry over $100 billion per annum, and the persistent problem of data governance, management, and analytics, it seems absurd to see financial institutions chasing will-o'-the-wisp solutions or technologies which may turn out to nothing more be fads, with little practical application or impact. Our previous research revealed that as of late 2016, the industry Chief Data Officers had yet to go beyond CDO 1.0 (Governance) to reach CDO 2.0 (Analytics) (Butler 2017). The point here being that the industry still finds difficulty in realising the benefits of data analytics, and this has major implications for RegTech, as it still has not solved the problems of data governance. Likewise, while it is clear that AI, machine learning, and robotics have significant implications for FinTech and, particularly, RegTech domains, the real benefits of AI, in terms of unsupervised learning, are still some way off. Nevertheless, it is clear that ontologies, machine learning, and NLP technologies are being used effectively. That said, it is clear that AI will do little to address the fundamental issue of "natural stupidity" in financial engineers and quants (Wilmott and Orrell 2017) or in financial experts and senior managers, whether in banks or general business organisations (Kahneman 2012), or in those responsible for systematic misconduct and fraud in the banking system (Vaughan and Finch 2016).

RegTech holds much promise for regulators and firms in the financial industry to fully benefit from the power that digitalisation offers—to solve a big problem for a relatively small effort. However, a considered, collaborative approach by all stakeholders is required if that promise is to become a reality. As societal stakeholders, IS researchers have a role to play in this, in that there is much the discipline can offer in helping practitioners develop and express a "common language" in human- and machine-readable formats.

Acknowledgements SmaRT was developed under the Enterprise Ireland Commercialisation Fund, co-funded by the Irish Government and European Regional Development Fund (ERDF) under Ireland's European Structural and Investment Funds Programme 2014–2020.

References

Arner, D. W., Barberis, J., & Buckey, R. P. (2016). FinTech, RegTech, and the reconceptualization of financial regulation. *Northwestern Journal of International Law & Business, 37,* 371–414.

Bamberger, K. A. (2009). Technologies of compliance: Risk and regulation in a digital age. *Texas Law Review, 88,* 669–739.

Butler, T. (2017). Towards a standards-based technology architecture for RegTech. *Journal of Financial Transformation, 45,* 49–59.

Haldane, A. G. (2012). *Towards a common financial language*. Securities Industry and Financial Markets Association (SIFMA) "Building a Global Legal Entity Identifier Framework" Symposium, New York. Available at: http://www.bis.org/review/r120315g.pdf. Last accessed 17 August 2018.

Heiler, S. (1995). Semantic interoperability. *ACM Computing Surveys (CSUR), 27*(2), 271–273.

JWG. (2017). *RegDelta: Part of our MiFID II solution*. Available at: https://jwg-it.eu/insight/mifid-programme-planner/. Last accessed 25 October 2017.

Kahneman, D. (2012). *Thinking, fast and slow*. England: Penguin Books.

Kontchakov, R., Rodriguez-Muro, M., & Zakharyaschev, M. (2013). Ontology-based data access with databases: A short course. In *Reasoning web: Semantic technologies for intelligent data access* (pp. 194–229). Berlin and Heidelberg: Springer.

Marenzi, O. (2017). *Capital markets and investment banking 2017–2018 forecast*. Available at: http://www.opimas.com/research/193/detail/. Last accessed 25 October 2017.

McDowell, H. (2017). Banks spent close to $100 billion on compliance last year. *The Trade News*. Available at: https://www.thetradenews.com/Sell-side/Banks-spent-close-to-$100-billion-on-compliance-last-year/. Last accessed 17 August 2018.

McNulty, L. (2017). Top regulator: City firms must bear responsibility for RegTech risks. *Financial News*. Available at: https://www.fnlondon.com/articles/city-firms-must-bear-responsibility-for-regtech-risk-20170516. Last accessed 25 October 2017.

Memminger, M., Baxter, M., & Lin, E. (2016). *Banking RegTechs to the rescue?* Available at: http://www.bain.com/publications/articles/banking-regtechs-to-the-rescue.aspx. Last accessed 17 August 2018.

Packin, N. G. (2018). RegTech, compliance and technology judgment rule. *Chicago-Kent Law Review, 93*, 193–220.

Palantir. (2018). *Our solutions*. Available at: https://www.palantir.com/solutions/. Last accessed 17 August 2018.

Palmer, J. (2017). *The data journey: Finding and fixing the bumps & holes in the road*. Data Standards for Granular Data Conference, European Central Bank.

Tyler, T. (2017). *RegTech and FinTech's impact on the regulated sector*. Available at: http://www.aidcompliance.com/regtech-fintechs-impact-regulated-sector/. Last accessed 17 August 2018.

Vaughan, L., & Finch, G. (2016). *The fix: How bankers lied, cheated and colluded to rig the world's most important number*. New York: Wiley.

Walker, O. (2018). M&A in asset management sector climbs to 8-year high. *The Financial Times*. Available at: https://www.ft.com/content/2f1e77f2-f80c-11e7-88f7-5465a6ce1a00. Last accessed 17 August 2018.

Walport, M. (2015). *FinTech futures: The UK as a world leader in financial technologies*. London: UK Government Office for Science.

Wilmott, P., & Orrell, D. (2017). *The money formula: Dodgy finance, pseudo science, and how mathematicians took over the markets*. Wiley.

Woolard, C. (2016). *Innovation in RegTech*. Available at: https://www.fca.org.uk/news/speeches/london-fintech-week-2016-innovation-regtech. Last accessed 17 August 2018.

CHAPTER 7

Payment Service Directive II and Its Implications

Alan Brener

Abstract The EU required member states to implement the new Payment Services Directive (PSD II) by January 2018. The European Banking Authority (EBA) will provide important final guidance on areas such as security during 2018, which will need to be implemented over the following couple of years. The increase in mobile and Internet banking and the failure of the original 2007 first Payment Services Directive (PSD I) to develop cross-border payment services encouraged the development of the revised Directive. The EU also took the opportunity to assist the development of new payment services, which may, in due course, disintermediate some of the traditional payment arrangements including, for example, those provided by credit card companies, and to reduce the cost of payments services for, primarily, businesses. It will pose challenges for banks and present opportunities for both new FinTech operations and large firms such as Apple and Amazon. The full

A. Brener (✉)
University College London, London, UK
e-mail: alan.brener@ucl.ac.uk

A. Brener
Queen Mary University London, London, UK

© The Author(s) 2019
T. Lynn et al. (eds.), *Disrupting Finance*, Palgrave Studies in Digital Business & Enabling Technologies,
https://doi.org/10.1007/978-3-030-02330-0_7

benefits of the new Directive will only be gained if a critical mass of customers see the value of the new services and trust the firms and processes involved.

Keywords Payment service · European Union · Single market · Customer protection · PSD II · FinTech

> All banks do is really data, so when you open that data up to third parties it allows for the first time a separation between the person that manages the customer relationship and the person that provides the balance sheet services. (Antony Jenkins, *Financial Times*, 12 January 2018)

7.1 Introduction

The new Payment Services Directive II (PSD II) is on the face of it, another technical piece of legislation. However, it is much more. It has been described as the EU firing the "starting gun for banks vs. fin-tech fight over payments" (Reuters 2017). It is both "another step towards a digital single market in the EU" and a move to introduce more competition into the EU's payments market and to break the banks' control over customer transaction information (Dombrovskis 2018).

A number of existing businesses may be disrupted by the developments encouraged by PSD II. These include credit card issuers and merchant acquirers, providing opportunities for new FinTech companies and very large firms such as Amazon, Apple, etc. There will be opportunities for firms that specialise in "account to account" transfers (A2A) and those who, for example, collect individual customer spending information, analyse the data and market it. Moreover, other jurisdictions are looking at EU legislative innovation which they may emulate (Yap 2017).

Payment services have largely avoided EU regulation until recently. However, regulation can "when drafted and applied correctly ... be an effective tool for creating incentives to increase innovation, economic development and competition" (Romānova et al. 2018, p. 21). This chapter looks at how the original view has changed with, initially, the first Payment Services Directive (PSD I); why PSD I was judged less than successful and the EU's attempt to get ahead of and, to a certain extent, guide the development of both markets and technologies which are fast changing through PSD II.

7.2   Background

In 2007 the EU published its first attempt at payment services regulation—the PSD I.[1] EU member states were required to implement the Directive in 2009. It was a maximum harmonisation Directive (i.e. EU states cannot exceed the terms of the Directive by, for example, imposing additional restrictions).

The central issue was that the payments systems within the EU were organised along national lines and fragmented. The aims of the Directive were to align these to help facilitate the EU single market in goods and services and to support greater competition in payment services (Donnelly 2016). Specifically, its objectives were to assist in the development of the Eurozone's cross-border payment system known as the Single Euro Payments Area (SEPA); to regulate payment businesses to encourage non-banks to enter the payments market; to increase services for customers by setting maximum payment processing times and standardised terms and conditions and to increase customer protection so that the latter would have greater confidence in the market.

Fundamental to this were provisions to ensure non-discrimination so that any payment service provider competing in the internal market could use "the services of the technical infrastructures" of incumbent payment systems providers on matching terms.[2]

The Directive was seminal, in that it set the foundations for future work to improve competition and innovation both within national jurisdictions and across the borders of EU states. It sought to break the associations of banks which, for example, in the UK had steered the payments systems. That it did not fully succeed is not to diminish the Directive's ground-breaking role as new technologies rapidly over-took legislation and existing market practices.

---

[1] 2007/64/EC, http://eur-lex.europa.eu/legal-content/EN/TXT/PDF/?uri=CELEX:32007L0064&from=EN.

[2] Ibid., PSD I, Recital 16.

7.3   EU Initiated Review of the Effectiveness of PSD I

The importance of the Directive is evidenced by the fact that relatively shortly after it came into effect, the EU organised an independent review ("the impact study") of its effectiveness.[3] The final report of the impact study prepared by London Economics and iff (in association with PaySys) was submitted in 2011. Its key findings addressed passporting, fees and charges for payment services, market fragmented and what are known as "one-leg" transactions (i.e. where funds are sent from an EU state to a non-EU jurisdiction). These issues are considered in more detail below.

The impact study praised the way the Directive had helped develop a single market in EU payment services and had increased transparency within the payments market and had also increased the speed at which they were executed. All this was seen as aiding business efficiency. No longer were electronic payments allowed to march at the speed of the slowest piece of paper through the payments' clearing system. However, there were still significant failures.

7.3.1   Main Findings of Impact Study

The impact study found little evidence of innovation in the market structure. There had been very few new entrants since the Directive came into force in 2009. Moreover, payment services firms had not grasped the opportunity to operate across EU borders using passporting privileges under the Directive.

PSD I required businesses offering payment services, whether within a single EU jurisdiction or across EU member state borders, to be authorised by their local or "home" state regulator. By late 2012 there were only 568 authorised payment institutions (APIs). Of these some 40% carried on the business of money remittance (i.e. sending money to non-EU states; often used by migrant workers). In spite of PSD I, there remained very wide differences between the structures of payment

[3] Study on the impact of Directive 2007/64/EC on payment services in the internal market and on the application of Regulation (EC) No. 924/2009 on cross-border payments in the Community, Final Report, http://ec.europa.eu/internal_market/payments/docs/framework/130724_study-impact-psd_en.pdf.

services providers across the various EU jurisdiction with no obvious explanations. 85% of the APIs existed before the Directive so there is no evidence of much new competition entering the market.

Additionally, the impact study found that the use of passporting for payment services varied greatly between jurisdictions but even when this legislative facility was employed, firms only operated in a small number of EU states besides their home nation. The process of obtaining a passport was seen as lengthy and complex. Reasons given for this included a lack of harmonisation of customer protection and anti-money laundering measures. The impact study also indicated that APIs that also provided credit to customers were subject to two separate regulators. It recommended that a single regulator supervise both the provision of credit and payment services.

One of the aims behind PSD I was to ensure equal charges for both domestic and cross-border payments within the EU for sums of €50,000 or less. However, the impact study found mixed results. In some instances, this had resulted in higher fees for both types of transaction and the introduction of new charges. Some EU states also permitted differential charges for different payment instruments reflecting the increased charges on merchants for credit card transactions. These charges could exceed the actual costs card companies imposed on merchants. This appears to contradict the Consumer Rights Directive.[4] This limits merchants charging "in respect of the use of a given means of payment, fees that exceed the cost borne by the trader for the use of such means".[5] However, the impact study did point out that establishing and enforcing the true cost to a merchant of accepting a credit card payment may be complex and difficult.

The impact study also found potential confusion between payments under the PSD I and those relating to e-money, which are subject to the Electronic Money Directive II.[6] In essence, a payment service provides secure messaging between the person or entity instructing the payment and the recipient of the funds and the respective businesses holding the money to be transmitted and the organisation receiving the funds.

---

[4]2011/83/EU, published in 2011 and enacted into national laws in 2013, https://eur-lex.europa.eu/legal-content/EN/TXT/?uri=CELEX%3A32011L0083.

[5]Ibid., Art 19.

[6]2009/110/EC, http://eur-lex.europa.eu/legal-content/en/ALL/?uri=CELEX:32009L0110.

The impact study considered that this process, and its importance, may not be clear to customers.

There are a number of payment services providers who were exempt from the Directive (e.g. pre-paid cards, ATM operators, money exchanges, etc.) whom the review, though, could be used to circumvent the Directive's requirements and hence gain an unfair competition advantage.

Another area of focus is known as "one-leg" transactions, mentioned earlier, since such transactions are normally undertaken by vulnerable migrant customers sending money home. The review recommended treating these types of transfers on the same basis as intra-EU payments. These and the other exemptions cause customer confusion since they may fail to understand which transactions are protected by the Directive and which fall outside it.

There was considerable confusion about the liability for unauthorised payments. Article 61 limited customer liability to €150 except in circumstances involving customer fraud or gross negligence. However, implementation in member states varied. The issue appeared to be the different evidential requirements demonstrating "gross negligence" in each jurisdiction.

Finally, the review reported large differences between national complaints arrangements required by the Directive. It praised those available in the Republic of Ireland and in the UK while observing that in most other member states, complaints systems had still to be developed.

In response to these findings in 2012, the EU Commission published a consultative "Green Paper": "Towards an integrated European market for card, Internet and mobile payments".[7] The Commission remained particularly keen to develop cross-border payments. However, it is possible to speculate that the Commission was also concerned that the major credit card companies continued to dominate the consumer payments system within the EU. This may be seen as reflected in the Commission's wish to help "to launch innovative, safe and easy-to-use digital payments services and to provide consumers and retailers with effective, convenient and secure payments methods in the Union".[8]

---

[7] http://ec.europa.eu/finance/consultations/2012/card-internet-mobile-payments/index_en.htm.

[8] PSD II, Recital 4.

7.4   Payment Services Directive II

In the light of this report and the rapid changes in technology, the EU quickly developed PSD II.[9] This repealed and replaced all the measures in PSD I. However, many articles in the original Directive were re-enacted in PSD II.

PSD II was published at the end of 2015 and required implementation in local law by January 2018. The Directive required the European Banking Authority (EBA) to develop a range of technical guidance to flesh out the Directive. These are considered later in this chapter.

The aims of the new Directive were to:

- assist in the integration of the EU's payments market,
- promote competition by encouraging new participants in the market including FinTech and the development of mobile and Internet payment services across the EU,
- encourage lower prices for payments, and
- increase customer confidence in making more efficient electronic payments by introducing better customer protection against fraud and other abuses and error. This would require enhanced security arrangements.[10]

The main themes in the Directive were to increase security measures and other customer protections, level the competitive playing field by reducing the various exemptions from payment services regulation and to permit two new innovative arrangements: "account information service providers (AISPs)" and "payment initiation service providers (PISPs)". These important developments are considered later. The next sections look at the other major changes first.

7.4.1   Scope of the Directive and the Removal of Exclusions

A number of exclusions exempting business operations from regulation have been removed. For example, payment arrangements which can only

---

[9] 2015/2366/EC, http://ec.europa.eu/finance/payments/framework/index_en.htm.

[10] European Commission—Fact Sheet, Payment Services Directive: frequently asked questions (12 January 2018), http://europa.eu/rapid/press-release_MEMO-15-5793_en.htm. Accessed 4 April 2018.

be used for buying goods and services from a prescribed list of businesses are now included within the Directive's scope.[11] However, payments made within a group of companies remains exempt from the need for regulation as do payments aimed at collecting funds for charitable purposes. As before, with PSD I, physical cash and paper based payment instruments (e.g. cheques) remain outside the scope of the Directive.

Payments sent or received where one of the Payment Service Providers (PSPs) is located outside the EEA will be covered, as will payments in non-EEA currencies.[12]

PSD II, as with PSD I, is limited to regulating payment services providers which do not also take deposits or issues electronic money. Firms which take deposits which are used to fund payments will continue to be regulated under the Capital Requirements Directive IV (i.e. banks and similar credit institutions).[13] Similarly, businesses which issue electronic money will continue to be subject to their own Directive.[14]

7.4.2   Authorisation of Payment Institutions

There are no substantial changes from PSD I on the authorisation and supervision of payment institutions. However, the EBA is tasked with the job of determining criteria for establishing the minimum amount of professional indemnity insurance or other forms of guarantee required by authorised firms. Moreover, the APIs will only be permitted to provide credit when it is closely linked to the payment service.[15]

In order to enhance co-operation between EU member states, the Directive requires the EBA to assist in resolving cross-border disputes between regulators and to publish guidance on this and the necessary data exchanges to aid supervision.[16] The EBA is also required to publish a central public register of authorised payment services firms.[17] The Directive contains various other customer protection measures such as

---

[11] PSD II, Recital 14.

[12] PSD II, Art 2.

[13] Directive 2013/36/EU.

[14] PSD II, Art 63 (3), The taking up, pursuit and prudential supervision of the business of electronic money institutions, Directive 2009/110/EC.

[15] PSD II, Art 1 (a).

[16] PSD II, Art 25 (5).

[17] PSD II, Art 15.

those relating to the transparency of charges and prohibitions on discrimination, based on nationality or place of residence against those residents legally in the EU.[18]

Host member states are permitted to take precautionary measures in the event of an emergency situation such as a large-scale fraud.[19]

7.4.3   Innovation

PSD II seeks to promote the development of two aspects of FinTech. The first collects, aggregates and analyses information from customer payments transactions. The Directive describes this as an "account information service" (AIS). PSD II views the second as a "software bridge between the website of the merchant and the online banking platform" of the customer initiating a payment across to the merchant's account.[20] It is classified in the Directive as a "payment initiation service" (PIS). It is defined in Article 4 (15) as "a service to initiate a payment order at the request of the payment service user with respect to a payment account held at another payment service provider". It is a secure messaging system and at no stage does the PIS provider ever hold the customer's payment.

Providers of such services are termed "PISPs" and "AISPs". They are also known collectively as third-party providers (TPPs).[21] These may be seen as distinct new financial services industries developing new customer services (Chiu 2017).

The Directive also refers to "account servicing payment service provider" (AS PSP). This is the firm where the customer's payment account is held (e.g. the customer's bank).

Customers must give explicit consent to use PIS and AIS arrangements. There is no requirement for a contract between the customer and either the PISP or AISP. Nor is a contract necessary between the PISP and the merchant supplying goods or services to the customer.[22] Customer agreements with PSP can be either ad hoc, good for a single transaction or set-up under a continuing contract. The latter must be

---

[18] PSD II, Art 98 and Title III.
[19] PSD II, Art 30.
[20] PSD II, Recital 26–29.
[21] PSD II, Art 4.
[22] PSD II, Recital 30.

capable of termination without charge with a notice period not exceeding a month.[23]

PISPs and AISPs must ensure that the personalised security credentials are not shared with other parties and they must not store sensitive payment data. AS PSPs are required to treat payment orders and data requests transmitted via a PISP or AISP "without any discrimination other than for objective reasons".[24]

However, both types of innovation enable third parties to delve into the payments accounts of customers. Hence the Directive delegates, to the EBA, the need to develop technical guidance for "secure customer authentication" (SCA). This important aspect is considered later.

7.4.4   Confirmation of Availability of Funds

PSD II creates a new fund availability confirmation service. It allows a third party with the customer's express permission to obtain confirmation from the customer's AS PSP (i.e. their bank) that sufficient funds are available to enable a payment to be made. It only requires a "yes/no" response.[25] It is not clear how useful this facility will be in practice since it is of little help in assessing credit worthiness. However, there may be some value in a merchant knowing that the funds exist to satisfy a payment a few moments before a payment order is executed on a customer's account.

7.4.5   Enhancing Competition

There is a broad requirement in the Directive that those participating in a payments system within the EU provide access to authorised payment services firms in a non-discriminatory way.[26] This is part of the general theme within PSD II promoting increased competition in payment services.

---

[23] PSD II, Art 55.

[24] PSD II, Recital 33.

[25] PSD II, Art 65.

[26] PSD II, Recital 50 and Art 69.

7.4.6   *Customer Protection*

Both the 2007 and the 2015 Directives on payment services are based on the understanding that in meeting their objectives customer trust is essential. PSD II, consequently, develops the protections provided initially by PSD I for individual "real" personal customers and EU member states are empowered to extend the Directive's safeguards to "micro-enterprises".[27]

Issues with incorrect or unauthorised payments should be communicated as soon as possible.[28]

There is an important protection afforded to customers in that the Directive requires that any alleged unauthorised transaction is immediately reimbursed unless there is a "high suspicion" that an "unauthorised transaction results from fraudulent behaviour" by the customer.[29] The suspicion must be based on "objective grounds". These must be passed to the national regulator and the PSP should "conduct, within a reasonable time, an investigation before refunding the payer".[30] Customers have eight weeks to make a claim for a refund.[31]

The customers, unless they are acting fraudulently or are grossly negligent, should only be liable for a maximum of €50 for any loss of their "payment instrument" (e.g. a payment access card) prior to their notifying the PSP.[32] What constitutes "gross negligence" will be a matter for national law. Any contractual attempt by a PSP to change or shift the burden of proof against the customer will be nugatory.[33]

The customer's PSP or PISP should assume responsibility for any failure in the payments chain.[34] However, if the customer has used the wrong payee's identifier, the PSP will not be liable but "should be obliged to cooperate in making reasonable efforts to recover the funds"

---

[27] Ibid., Art 4 (36).
[28] PSD II, Arts 73–74.
[29] Ibid.
[30] Ibid.
[31] PSD II, Art 77.
[32] PSD II, Art 74.
[33] PSD II, Recital 72.
[34] PSD II, Art 90.

including providing information to the customer to help trace the missing funds.[35]

In terms of liability, in the event of an unauthorised, non-executed, defective or late executed payment initiated via a PISP, the AS PSP is required to refund the customer immediately. There is an obligation on the PISP to compensate the AS PSP where the former is liable, with the burden of proof lying with the PISP "to prove that, within its sphere of competence, the payment was authenticated, accurately recorded and not affected by a technical breakdown or other deficiency," linked to the payment service of which it is in charge.[36]

The Directive stipulates that the full amount transferred should arrive intact without any charges being levied beyond those agreed at the outset.[37]

All payment made in Euros or other member state currencies should be executed within, at most one day. All other payments should also be completed within the same time period unless otherwise agreed.[38]

7.4.7   Security

Security measures must be proportionate to the security risk and PSPs must maintain measures to mitigate security risks and to provide the national regulator with regular updates assessing these risks together with their risk reduction actions.[39] PSPs are under an obligation to report, quickly, major security incidents to national authorities.[40]

7.4.8   Complaints Handling

The Directive requires that member states have an easily accessible, independent, impartial, transparent and effective alternative disputes resolution arrangement for issues between customers and PSPs.[41] PSPs must

---

[35] PSD II, Recital 88.
[36] PSD II, Art 72.
[37] PSD II, Art 81.
[38] PSD II, Art 83.
[39] PSD II, Recital 91.
[40] PSD II, Art 96.
[41] PSD II, Art 102.

have dispute resolution procedures and must respond to complaints within fifteen business days of a complaint being received.[42]

7.5   EUROPEAN BANKING AUTHORITY (EBA) WORK ON PSD II

The EBA has a series of work projects in-hand on the implementation of PSD II to ensure that they are secure and efficient.[43] It has been preparing a Regulatory Technical Standard (RTS) on home/host state cooperation and, in particular, the information exchanges needed by both. This includes separate guidance on the reporting of fraud by PSPs to local competent authorities.

The EBA has also produced an RTS and a set of Implementing Technical Standard (ITS) on setting up the EBA register mentioned earlier. There is also guidance on areas such as professional indemnity insurance. Important technical guidance on security measures and SCA, incident reporting and complaints handling have been agreed and published. SCA is considered in more detail later below (see also Zetzsche et al. 2017).[44]

7.6   SECURE CUSTOMER AUTHENTICATION (SCA)

As part of the move to protect customers and businesses, PSD II requires SCA—which authenticates the identity of the customer and their right to make the transaction—before an electronic payment can be made.[45] SCA "is based on the use of two or more elements categorised as knowledge

[42] PSD II, Art 101.

[43] https://www.eba.europa.eu/regulation-and-policy/payment-services-and-electronic-money/-/activity-list/MgjX6aveTl7v/more. Accessed 9 April 2018. See also 'EBA mandates in PSD2 and their timelines', https://www.eba.europa.eu/documents/10180/87703/EBA+Mandates+PSD2.pdf/5c2493a4-ef26-4434-8338-736895bd423f

[44] The EBA has stated that it will be "analysing regulatory sandboxes [safe regulatory areas for testing innovative products, services and operations] and innovation hubs with a view to developing a set of best practices to enhance consistency and facilitate supervisory coordination", EBA FinTech Roadmap (March 2018), 4, https://www.eba.europa.eu/documents/10180/1919160/EBA+FinTech+Roadmap.pdf.

[45] Supplementing Directive 2015/2366 of the European Parliament and of the Council with regard to RTS for SCA and common and secure open standards of communication, http://ec.europa.eu/finance/docs/level-2-measures/psd2-rts-2017-7782_en.pdf.

(something only the user knows, e.g. a password or a PIN), possession (something only the user possesses, e.g. the card or an authentication code generating device) and inherence (something the user is, e.g. the use of a fingerprint or voice recognition)".[46] There is a view that these arrangements may "ring alarm bells" as these services "open up a new class of vulnerabilities" (Mansfield-Devin 2016). "For remote transactions, such as online payments, the security requirements go even further, requiring a dynamic link to the amount of the transaction and the account of the payee, to further protect the user by minimising the risks in case of mistakes or fraudulent attacks".[47]

7.6.1   *Exemptions for SCA*

"As a matter of principle, all electronic means of payment are subject to the requirement for SCA. However, exemptions are possible as it is not always necessary and convenient to request the same level of security from all payment transactions".[48] For example, low value transactions such as that used for contactless payments at terminals should not require SCA.[49]

7.7   COMMENTARY

It is not immediately obvious how the availability of PISs will change how customers operate. Customers will not see much change if they use a PISP compared to using their current credit or debit card for making a payment. However, credit card issuers and acquirers are likely to be disintermediated since merchants will not need their services. The PISP will move the funds straight from the customer's bank account into that of the merchant.

It is likely that this will be cheaper for merchants who, in any event, are not permitted to charge extra for different payment methods under the Directive (Grüschow et al. 2016). It may be possible for the merchant to pass some of the margin saved to the customer but again how this might be done is still not clear since offering a discount, say, for those using a PIS compared with a credit card would fall foul of PSD II.

---

[46] European Commission—Fact Sheet—PSD II: frequently asked questions, 16, http://europa.eu/rapid/press-release_MEMO-15-5793_en.htm. Accessed 4 April 2018.

[47] Ibid., 16.

[48] Ibid., 17.

[49] Ibid., 17.

AISPs may be able to help customers who have multiple financial products which they want to view regularly. With the customer's express permission, the AISP could access all the customer's accounts in the EU and present the information in near enough real time. The data could be expressed in charts and analysed into different categories of expenditure.

This information would be of value in the market both in aggregate and individually. It would help firms decide what to market and to whom. It would be of value to competitors since, for example, a customer could be enticed to move their current account with a cheaper overdraft offer.

However, it is not clear what actual level of customer demand exists for AIS. Typically, in the EU, only around 19% of customers have more than two bank accounts (EY 2012). Further, where a customer has two bank accounts, one will normally be for their banking transactions and the other for savings. There is a view that in Europe the advantages for customers of A2A have yet to emerge (Wyman 2016). Banks will almost certainly act to protect their current positions since it is estimated that some 9% of retail payments revenue may be under threat by 2020 (Jackson 2018). The evidence is that most customers are very passive; reluctant to change "their" bank and it usually takes a significant operational failure to prompt a customer to move accounts (European Commission 2007).

There is scope for future socio-legal research on both merchant suppliers and customer attitudes to the changes brought about by PSD II. Various businesses will be undertaking their own research but they are unlikely to approach it from the legal perspective. The EU will probably review whether the results from the Directive demonstrate that the markets in payment services are moving towards meeting its own objectives. Indeed, the EU will need to keep this whole area under close review as a result of both social and technological changes affects the markets and customer outcomes. Much also will depend on fraud prevention where even SCA may prove vulnerable (European Payments Council 2017).

PSD II provides scope for FinTech to develop in key parts of the payment services market. However, it is likely that growing market share will be a significant challenge for small innovators. Nevertheless, there are opportunities for large players such as the Apples and Amazons' of this world to gain margin from card companies and for banks to introduce their own A2A arrangements buttressed by their reputation with customers.

REFERENCES

Chiu, I. H. (2017). A new era in FinTech payment innovations? A perspective from the institutions and regulation of payment systems. *Law, Innovation and Technology, 9*(2), 190–234.

Dombrovskis, V. (2018). *Payment services: Consumers to benefit from cheaper, safer and more innovative electronic payments.* Available at: http://europa.eu/rapid/press-release_IP-18-141_en.htm. Last accessed 4 April 2018.

Donnelly, M. (2016). Payments in the digital market: Evaluating the contribution of Payment Services Directive II. *Computer Law & Security Review, 32*(6), 827–839.

European Commission. (2007). *Commission report of the expert group on customer mobility in relation to bank accounts.* Available at: http://ec.europa.eu/internal_market/finservices-retail/docs/baeg/report_en.pdf. Last accessed 16 August 2018.

European Payments Council. (2017). *2017 payment threats and fraud trends.* Available at: https://www.europeanpaymentscouncil.eu/sites/default/files/kb/file/2017-12/EPC214-17v1.0%202017%20Payment%20Threats%20and%20Fraud%20Trends%20Report_1.pdf. Last accessed 16 August 2018.

EY. (2012). *The customer takes control: Global Consumer Banking Survey 2012.* Available at: http://www.ey.com/Publication/vwLUAssets/ey-global-consumer-banking-survey-2012/$FILE/ey-global-consumer-banking-survey-2012.pdf. Last accessed 16 August 2018.

Grüschow, R. M., Kemper, J., & Brettel, M. (2016). How do different payment methods deliver cost and credit efficiency in electronic commerce? *Electronic Commerce Research and Applications, 18*, 27–36.

Jackson, O. (2018). PSD2 gives banks chance to evolve. *International Financial Law Review.*

Mansfield-Devine, S. (2016). Open banking: Opportunity and danger. *Computer Fraud & Security, 2016*(10), 8–13.

Romānova, I., Grima, S., Spiteri, J., & Kudinska, M. (2018). The Payment Services Directive 2 and competitiveness: The perspective of European FinTech Companies. *European Research Studies Journal, 21*(2), 5–24.

Thomson Reuters. (2017). *US and EU payments regulation: EU fires starting gun.* Available at: https://www.reuters.com/article/us-eu-payments-regulations/eu-fires-starting-gun-for-banks-vs-fintech-fight-over-payments-idUSKBN1DR1AZ. Last accessed 4 April 2018.

Wyman, O. (2016). *EU retail and SME payments: The state of the industry.* Available at: http://www.oliverwyman.com/content/dam/oliver-wyman/v2/publications/2016/Nov/European-Retail-and-SME-Payments-web.pdf. Last accessed 16 August 2018.

Yap, B. (2017, February 17). New EU payment services rules spur new regulation in Japan. *International Financial Law Review*, 1–4.

Zetzsche, D. A., Buckley, R. P., Barberis, J. N., & Arner, D. W. (2017). Regulating a revolution: From regulatory sandboxes to smart regulation. *Fordham Journal Corporate & Financial Law, 23*, 31.

CHAPTER 8

From Transactions to Interactions: Social Considerations for Digital Money

Jennifer Ferreira and Mark Perry

Abstract In our highly connected world, the number of digital transactions is growing, and so too are the myriad of digital platforms that enable these transactions. While the dominant perspective on developing digital payment platforms involves implementing an efficient, low cost, and secure transfer of value, in this chapter, we take a step back to re-examine how digital transactions are embedded in social relationships, and that by focusing solely on the transfer of value, it is possible to miss opportunities for social interactions in digital transactions. We examine the affordances of digital transactions to illustrate possibilities for action, opportunities for interaction, and the roles of negotiation and intermediation within digital transactions. We then highlight some social impacts of digital transactions and its associated data generation, its embeddedness alongside other available forms of transaction, and the ways in which the digital world conflates money with payment systems.

J. Ferreira (✉)
University College Cork, Cork, Ireland
e-mail: jferreira@gmx.com

M. Perry
Brunel University London, London, UK
e-mail: mark.perry@brunel.ac.uk

© The Author(s) 2019
T. Lynn et al. (eds.), *Disrupting Finance*, Palgrave Studies
in Digital Business & Enabling Technologies,
https://doi.org/10.1007/978-3-030-02330-0_8

121

Keywords Digital money · Financial affordance · Social transactions · Interaction design · User experience design

8.1 Introduction

In standard economic texts, money is usually referred to as a unit of account, a store of value, and a medium of exchange (Asmundson and Oner 2012), and yet when examined as a social phenomenon, other critically important attributes about our societies are revealed, most notably, how trust (Ingham 2004) and power structures (Baker 1987) are operationalised. This view that money's extraeconomic, social basis (Zelizer 2011) should be acknowledged is one that carries increasing weight: the usefulness and value of money—and its concomitant forms of exchange—are socially constructed and locally contingent. Indeed, Simmel's classic text on the philosophy of money (Simmel 1900), in which he examines the mechanisms that underpin economic exchange, considers financial transactions as a form of social interaction and adds that outside the exchange relation, money loses meaning. Dodd upholds a similar view of the inseparability of money and social relations: "[...] money is a process, not a thing, whose value derives from the dynamic, ever-changing, and often contested social relations that sustain its circulation" (Dodd 2014, Preface). This view allows us to move beyond the abstract, a socialised flow of value that are typical of the literature in economics (Zelizer 2011).

While we are seeing a resurgence in how the use of money is understood, at a practical level, this is being challenged by the move towards making money digital. For technology developers and the banking industry, money appears to be envisaged simply as aseptic, standardised data in binary form, viewed as an online resource, with payment promoted as an efficient form of value token transaction as it moves across digital networks and is audited via a remote banking ledger (Wandhöfer 2017). We argue that this view is not wrong, but that it is a very partial perspective. Nevertheless, it is worth summarising the reasons for this orientation towards money as data so that we can explore its drivers and then begin to dissect how this may limit a discourse about its design and use. It is evident that connected ubiquitous mobile technologies have opened up opportunities for innovative financial technology solutions for storing money and payment instruments (e.g. digital wallets), and conducting transactions (e.g. Apple Pay, Square, Stripe). The benefits of transacting digitally, in large part lies in the speed with which identities can be verified and transactions confirmed. Digital payments (see Chapter 7) are considered to be a cheaper form of transaction, and it is estimated that the use of cash can cost countries more than 1% of their Gross Domestic Product (GDP) (Denecker et al. 2013); electronic payment systems have lower administrative costs, lower security costs, and digital money does not require transportation, so that such reduced digital infrastructural overheads offer considerable advantages. So there is a reasonable argument to be made for considering the value of financial digitalisation as presenting efficiency gains with faster, cheaper, and more mobile transactions. Yet, as we have argued, this ignores aspects of money that are pervasive even in quantitative fields such as economics in which the use of money (through prices) provides information about markets, so that money necessarily involves *inter*actions, not just *trans*actions. So how do we account for the social aspects around transactions? Returning to the conceptualisation of financial transactions embedded in social relations, we investigate where opportunities are for social interactions in digital transactions.

In this chapter, we examine the affordances of digital transactions to illustrate possibilities for action, opportunities for interaction, and the roles of negotiation and intermediation within digital transactions. We then highlight some social impacts of digital transactions and its associated data generation, its embeddedness alongside other available forms of transaction, and the ways in which the digital world conflates money with payment systems.

8.2   Affordances of Digital Money

How money is designed lends itself to different forms of use, both physically and socially, and we refer to these forms of use as *affordances*. Norman (1999, p. 39) explains that "the word affordance was coined by the perceptual psychologist J. J. Gibson (1977) to refer to the actionable properties between the world and an actor (a person or animal)." Similarly, how money is represented shapes the ways in which we can interact with it and use it. Different materials offer different affordances and 'forcing functions' (i.e. constraints on use; Norman 1999), historically illustrated by Jevons from classical literature, noting "there was a tradition in Greece that Lycurgus obliged the Lacedæmonians to use iron money, in order that its weight might deter them from overmuch trading" (Jevons 1876, p. 53). These notions of portability designed into the currency would seem to have had a direct impact on use, much in the same way that certain configurations of digital money and payment operations might confer financial benefits around Anti Money Laundering (AML) or Know Your Customer (KYC) regulations, or tracking data on user spending or income might, for example, allow different forms of social sharing, customised marketing, preferential interest rates, or personalised financial services to be made available to users. Here we list some of the affordances of digital money and the opportunities for digital value transfer:

Frictionless: Digital money offers the promise of frictionless transactions. That is, financial interactions that are fast and easy, enabled by contactless technologies. For example, checkout terminals supporting contactless payments are becoming increasingly ubiquitous, where payment simply requires the consumer to wave a card or device in front of a reader, eliminating the need for entering a PIN or swiping a card through the machine. Likewise, but offering a different set of social affordances, Alipay in China supports the use of personalised QR codes, where payment is completed with a scan of the QR code. But even these forms of directly replacing the traditional form of payment are being challenged as digital technology can reformat the nature of the transaction dramatically. Thus, using a combination of computer vision, sensor fusion, and deep learning with a mobile app, Amazon Go stores eliminate the need for consumers to pass through a checkout point altogether, and therefore eliminate the need for any interaction on the part of the consumer. Described as 'grab-and-go,' the transaction is frictionless in that it is fully automated—consumers enter the store, select their items and then leave. Giving users a means of making sense of how this works, when it is operational, and which payment system is currently operating will be a significant challenge in order to ensure user understanding and trust.

Anonymous: Digital money offers new possibilities for conducting transactions anonymously. Persisting weaknesses in Internet security and privacy concerns drive the need for technological solutions that protect consumers' identities (Juang 2003). One early example of an attempt at an anonymous payment system was e-cash, invented by David Chaum, as a type of limited-traceability system (Chaum 1983; Chaum and Brands 1997). The aim was to emulate the anonymity of cash transactions, through cryptographic protocols (Goldberg et al. 1997).

Bitcoin, currently the most widely known cryptocurrency, is often cited as an anonymous currency, but it is in fact pseudonymous (Anonymous 2017). Full anonymity, requires hiding not only the identities of those involved in the transaction, but also the content of the transaction as well as metadata such as the date of transaction and method of payment. Further, anonymity in transactions tends to be traded off against speed and requires high levels of processing power. Achieving fully anonymous digital transactions is still an ongoing challenge, and the socio-political value that payment anonymity holds is a topic of contentious debate. Anonymous—and partially anonymous—payments can be problematic in the context of customer service: showing that something has been paid for by a customer (e.g. for item pickup or returns) when it is unclear who has paid is likely to present difficulties when these scenarios are not actively considered by designers.

Transparent: Digital money offers mechanisms for transparency in financial transactions. Blockchain technology (see Chapter 10), popularised by Bitcoin "[..] offers a way of recording transactions or any digital interaction in a way that is designed to be secure, transparent, highly resistant to outages, auditable, and efficient" (Schatsky and Muraskin 2015). Transparency around transactions allows auditing, gauging who you are transacting with, and can help build trust and discourage fraudulent transactions. While this enables an "unprecedented level of forensic analysis to be carried out on the transactions themselves" (Buenaventura 2017, p. 26), it also allows the transactional metadata to be used by other parties, which might include banks, third-party financial services, government agencies and tax authorities, or even users themselves in exploring their patterns of spending. Herein lies a challenge for designers in determining how transparency is managed, and who has access to what information. To what extent would you like others to know your financial arrangements in the same way that Google, Facebook, and Amazon know about your digital existence: Your partner? Your boss? Your bank? Your life or health insurance company? The government? Permissioned transparency also potentially offers criminals access through a backdoor to users' financial records.

Non-denominated: Digital money is divisible in ways that physical money is not. The use of digital money allows micropayments; payments that may even be below normal minimum denominations of currency (e.g. sub-cent or sub-penny). This is made more plausible when transaction charges are low, which enable payments in very small amounts to be made viable. Early instantiations of micropayment systems faltered in the 1990s, but blockchain technology, with its potential for low-transaction cost micropayments offers credible opportunities for casual and ad hoc payments. This has been demonstrated as having value in thing-to-thing (also known as machine-to-machine) payments in the Internet of things (Lundqvist et al. 2017), for example, to purchase or sell power, bandwidth, or data. The effort of making or setting up multiple tiny payments manually, is however, a challenge, and allowing end users (i.e. ordinary citizens) to set these payments up, to monitor them over time, and to ensure that fraudulent payments are not being made requires user interface designs that are easily understandable and simple to operate.

Dataful: Using digital money itself generates data, in a way that using cash does not. This data has enormous potential value and can be used to both generate revenue with new business models, as well as to provide the users themselves with information about their monetary activities, in the same way that Google gains knowledge through users' search activities at the same time as users can gain access to more personalised knowledge as they do so.

8.3   Opportunities for Interaction

For money use to be conceived as social *inter*actions, rather than just *trans*actions, hinges on identifying the opportunities available for money users to engage in social encounters with each other. In the following sections, we draw attention to where these opportunities might be in the transaction, the effects of intermediation on these opportunities, and implications for understanding value in the transaction.

8.3.1   Negotiating Payment

Two parties coming to agreement on how payment will proceed, what information will be exchanged and how. While a typical cash transaction occurs during a face-to-face exchange of cash, it is easy to imagine any means of exchange using similar physical formats of money. As one example, one party could place their money in a physical location and hand the other party a set of instructions for how to locate it. The mechanics we choose to adhere to during the exchange of money in our everyday lives are guided by social conventions (Carruthers 2010), but

not limited to them. Consequently, the rules for the exchange of physical money can be considered *negotiable* by the transacting parties and,
hence, become opportunities for interaction.

8.3.2 Effects of Intermediation

The more the transaction is intermediated (by banks, financial institutions, technology, and infrastructure companies), the less choice transacting parties have in setting the rules of the value transfer. For example, a
payment involving a bank deposit, will heed the rules as set by the bank
and the regulatory framework in which the bank operates. Negotiable
matters between the transacting parties are mostly limited to non-
procedural decisions such as the agreement that bank deposit money is
a valid form of payment and which banks may be involved. When payments involve digital money, the tools used in the transfer of digital
money, and by implication the designers of those tools, further constrain
which decisions are left to transacting parties concerning the rules of the
value transfer.

8.3.3 Collaborative Value Creation

During a transaction, the value is not only something that is transferred
between transactors, it can also be created by virtue of the interaction
between the transacting parties (Carroll and Bellotti 2015). When people (and devices) have to work together, that is, purposefully coordinate
their actions to accomplish a monetary transaction, these transactors are
engaging in what has been found to be a valuable set of opportunities for
building social connections (Ferreira et al. 2015). In this way, transactors are creating value in the exchange that extends beyond its economic
value. Research into cumbersome transactions, that is, transactions that
are perceived as slow or tedious, has highlighted the ways in which people engage with each other during the transaction and the implications of
this type of interaction for enriching their social relationships. Bringing
digital payment devices, such as mobile phones, into the exchange sets
up the interactions with yet more potential variations.

What this would suggest is that the payments technology used—infrastructure, interaction design, and physical form factor—offer ways of creating new connections between people and new ways of using money to
drive social interactions.

8.4   Social Impacts of Digital Transactions

Along with recognising the opportunities for social encounters in transactions, there is a need to examine the social impacts brought on by transacting digitally. In this section, we present our observations on social impacts with respect to attitudes around financial data, the availability of different forms of money, and understanding of money and payment systems.

8.4.1   Sensitive Data Generation and Sharing

Financial and credit card data have been shown to be considered the most sensitive personal data (Rose et al. 2013) and Experian cyber analysts estimate the value of stolen financial data to be worth up to USD 200 on the Dark Web (Stack 2018). In order to protect financial data, laws and regulations have emerged that impose strict security requirements on the institutions and other financial service providers that process financial data. As a result, there are limitations on how financial data can be shared or opened up for inspection. In Europe, there have been moves to allowing third parties (in practice, new FinTech entrants being allowed banking) permissioned access. This poses challenges for new entrants to the FinTech space where no or limited access to data requires creative workarounds in the design of technologies that interface with financial infrastructures.

8.4.2   Choice Proliferation

Despite the drive towards cashless societies, digital money and digital exchange continue to co-exist alongside non-digital forms of money and non-digital forms of exchange. Increasingly, the money we use and the ways in which it is exchanged are understood to be a collection of pragmatic responses to wide-ranging needs. So despite governmental and regulatory attempts at homogenising the money system, the varieties of uses and social contexts that emerge around money continue to engender new forms of money and exchange. For example, loyalty points, and volunteer currencies such as time dollars. The proliferation of connected digital tools that enable new forms.

An integrated approach to parallel physical and digital media seems to be a prevalent concern across fiat and alternative currencies (see, for example, the Bristol and Brixton Pounds—Perry and Ferreira 2018). Similarly, O'Neill et al. (2017) discuss the challenges and user practices around working digital money into the cash economy; these are non-trivial problems for users in making money work for them, in their individual and local circumstances. Moreover, the physical work that goes into making digital transactions also presents a considerable challenge to designers as they look to develop useful affordances into digital forms of money.

8.4.3   Untangling Money and Payment System

When we talk about digital payment systems, or refer to digital or mobile money, we refer not to money as an object of value itself, but to our use of the digital infrastructure that has been built up around the intermediated transfer of value using bank deposit money. This, along with the limited scope for ordinary people to negotiate their own rules around digital payment systems, invites a confusing situation where 'money' and 'payment system' become increasingly difficult concepts to separate in the digital world. In the world of cryptocurrencies, this is taken a step further, with the payment system (as a digitally held ledger of balances) wholly superseding the need for an object—we might call this money—itself. Here, as with a card payment or bank transfer, no actual thing is transferred, but merely a digital record is updated. In this respect, the payment system—the financial infrastructure—takes on the role of money. However, this is very much at odds with the ways that most everyday users of money tend to conceive of its operation when they make or receive payments. The phrase "I'll pay you" is very much an active process of transfer, compared to the reality of transactional settlement that might be functionally better expressed as "I'll initiate a permissioned record change to your bank account." Moreover, there is often very little directness to this financial transfer, with a variety of intermediaries sitting between payer and payee, to the extent that actual financial settlements between the payer and payee's banks may only happen at a single point as an aggregate of all customer transactions between institutions over the accounting time period. When designing payment systems for customers, it may therefore be necessary to represent what actually happens in ways that map more to user perceptions of this process than to institutional actuality.

8.5   Conclusion

Money is a multifaceted dynamic concept and our understanding of it is continually challenged and modified by financial innovations. Perhaps due to a long tradition within monetary theory of treating money as 'neutral,' there is still no agreed position on what money is and, despite ongoing critique, the 'textbook triad' of money as a unit of account, a store of value and a medium of exchange, continues to structure much of the discussion. This difficulty carries over to discussions of digital money and digital transactions. Understanding its use and working to change its operation is an especially complex task precisely because money is so pervasively connected to our lives.

In this chapter, we examined the affordances of digital transactions to illustrate possibilities for action, opportunities for interaction, and the roles of negotiation and intermediation within digital transactions. We have highlighted some social impacts of digital transactions and its associated data generation, its embeddedness alongside other available forms of transaction, and the ways in which the digital world conflate money and payment systems.

As digital money plays an increasingly central role in our lives, having the means to articulate our interactions with it helps to ensure digital transactions are designed to be the kinds of experiences we wish to have. Payment platforms, like any other digital tools, are open to be shaped by their designers, and can do more to support the interactional and transactional work required—future systems that attend to their users' needs offer opportunities that extend far beyond the rather limited current notions of faster payments and cheaper services.

References

Anonymous. (2017, August 23). Bitcoin transactions aren't as anonymous as everyone hoped. *MIT Technology Review*. Available at: https://www. technologyreview.com/s/608716/bitcoin-transactions-arent-as-anonymous-as-everyone-hoped. Last accessed 19 July 2018.

Asmundson, I., & Oner, C. (2012). What is money? *Finance & Development, 49*(3). Available at: http://www.imf.org/external/pubs/ft/fandd/2012/09/basics.htm. Last accessed 14 May 2018.

Baker, W. E. (1987). What is money? A social structural interpretation. In M. S. Mizruchi & M. Schwartz (Eds.), *Intercorporate relations* (pp. 109–144). Cambridge, MA: Cambridge University Press.

Buenaventura, L. (2017). *Reinventing remittances with Bitcoin.* Singapore: Bloom Solutions.

Carroll, J. M., & Bellotti, V. (2015). Creating value together: The emerging design space of peer-to-peer currency and exchange. In *Proceedings of the 18th ACM Conference on Computer Supported Cooperative Work & Social Computing (CSCW '15)* (pp. 1500–1510). New York, NY: ACM.

Carruthers, B. G. (2010). The meanings of money: A sociological perspective. *Theoretical Inquiries in Law, 11*(1), 51–74.

Chaum, D. (1983). Blind signatures for untraceable payments. In D. Chaum, R. L. Rivest, & A. T. Sherman (Eds.), *Advances in cryptology.* Boston, MA: Springer.

Chaum, D., & Brands, S. (1997). Minting' electronic cash. *IEEE Spectrum, 34*(2), 30–34.

Denecker, O., Istace, F., & Niederkorn, M. (2013). Forging a path to payments digitization. *McKinsey on Payments, 16,* 3–9. Available at: https://www.mckinsey.com/~/media/mckinsey/dotcom/client_service/financial%20services/latest%20thinking/payments/mop16_forging_a_path_to_payments_digitization.ashx. Last accessed 30 July 2018.

Dodd, N. (2014). *The social life of money.* Princeton: Princeton University Press.

Ferreira, J., Perry, M., & Subramanian, S. (2015). Spending time with money: From shared values to social connectivity. In *Proceedings of the 18th ACM Conference on Computer Supported Cooperative Work & Social Computing (CSCW '15)* (pp. 1222–1234). New York, NY: ACM.

Gibson, J. J. (1977). The theory of affordances. In R. E. Shaw & J. Bransford (Eds.), *Perceiving, acting, and knowing.* Hillsdale, NJ: Lawrence Erlbaum Associates.

Goldberg, I., Wagner, D., & Brewer, E. (1997). Privacy-enhancing technologies for the internet. In *Proceedings IEEE COMPCON 97* (pp. 103–109). San Jose, CA: Digest of Papers.

Ingham, G. (2004). *The nature of money* (1st ed.). Cambridge, UK: Polity Press.

Jevons, W. S. (1876). *Money and the mechanism of exchange.* New York: D. Appleton and Co. Available at: http://www.econlib.org/library/YPDBooks/Jevons/jvnMMECover.html. Last accessed 14 May 2018.

Juang, W.-S. (2003). A practical anonymous payment scheme for electronic commerce. *Computers & Mathematics with Applications, 46*(12), 1787–1798.

Lundqvist, T., de Blanche, A., & Andersson, H. R. H. (2017). *Thing-to-thing electricity micro payments using blockchain technology* (pp. 1–6). 2017 Global Internet of Things Summit (GIoTS), Geneva.

Norman, D. (1999). Affordance, conventions, and design. *Interactions, 6*(3), 38–41.

O'Neill, J., Dhareshwar, A., & Muralidhar, S. H. (2017). Working digital money into a cash economy: The collaborative work of loan payment. *Computer Supported Cooperative Work, 26*(4–6), 733–768.

Perry, M., & Ferreira, J. (2018). Moneywork: Practices of use and social interaction around digital and analog money. *ACM Transactions on Computer-Human Interaction (TOCHI), 24*(6), 41:1–41:32.

Rose, J., Barton, C., Souza, R., & Platt, J. (2013). *The trust advantage: How to win with big data.* Available at: www.bcgperspectives.com/content/articles/ information_technology_strategy_consumer_products_trust_advantage_win_ big_data/. Last accessed 14 May 2018.

Schatsky, D., & Muraskin, C. (2015). *Beyond Bitcoin: Blockchain is coming to disrupt your industry.* Deloitte University Press. Available at: https://www2. deloitte.com/insights/us/en/focus/signals-for-strategists/trends-block-chain-bitcoin-security-transparency.html. Last accessed 28 August 2018.

Simmel, G. (1900). *A chapter in the philosophy of value* (3rd ed., D. Frisby, Ed., T. Bottomore & D. Frisby, Trans.). Routledge: London.

Stack, B. (2018). Here's how much your personal information is selling for on the dark web. *Experian.* Available at: https://www.experian.com/blogs/ ask-experian/heres-how-much-your-personal-information-is-selling-for-on-the-dark-web/. Last accessed 30 July 2018.

Wandhöfer, R. (2017). The future of digital retail payments in Europe: A role for central bank issued crypto cash? Digital transformation of the retail payments ecosystem. *ECB and Banca d'Italia Joint Conference,* 30 November and 1 December 2017, Rome, Italy. Available at: https://www.ecb.europa. eu/pub/conferences/shared/pdf/20171130_ECB_BdI_conference/pay-ments_conference_2017_academic_paper_wandhoefer.pdf. Last accessed 13 May 2018.

Zelizer, V. (2011). *Economic lives.* Princeton: Princeton University Press.

CHAPTER 9

Token-Based Business Models

Paolo Tasca

Abstract Crypto assets can be classified into two main categories, according to their principal function: native coins and crypto tokens. Native coins, like Bitcoin, generally compete with the traditional forms of money providing both an alternative currency instrument and a payment infrastructure. Differently from native coins, crypto tokens are coins that embed some intrinsic values somehow linked to the quality of the issuing entity's business model and to the ecosystem it generates. This chapter explores the emergent start-up token funding model of Initial Coin Offering, which allows entrepreneurs to bypass the traditional capital market by issuing crypto tokens directly to investors. The positive feedback loop between an issuer's business model and the token funding model will be demystified.

Keywords Blockchain · Bitcoin · Cryptocurrency · Token · Platform business model · Fund-raise · Initial coin offering

P. Tasca (✉)
Centre for Blockchain Technologies,
University College London, London, UK
e-mail: P.Tasca@ucl.ac.uk

9.1   INTRODUCTION

Money is a social invention (Samuelson 1958; Menger 1892). A cryptocurrency is "a digital currency in which encryption techniques are used to regulate the generation of units of currency and verify the transfer of funds, operating independently of a central bank" (Oxford Dictionary). Nowadays, there exist several cryptocurrencies, more than one thousand, and in the years to come tens of thousands cryptocurrencies are expected to populate our economy in a sort of currency competition à la Hayek.

Most of these cryptocurrencies have a public common underlying ledger termed blockchain (see Chapter 10), where tamper-proof blocks of transactions are linked through an "append-only" logic following a predefined set of rules. The structure has been engineered in order to allow users to trust the process, not the counterparty. Thus, users can exchange valuable information or monetary value even without knowing each other, their geographical position, their affiliation or nationality and especially their reliability. For the sake of simplicity, we refer to blockchain as the larger family of distributed ledger technologies, which encompass also non-block-based ledgers (e.g., Ripple or IOTA).

Cryptocurrencies represent the latest step of technology evolution in terms of currencies: a long process that has unfolded through millenniums of trading from barter to the dematerialisation of banknotes that is bringing us to digital fiat. The recent development of peer-to-peer (P2P) networks, the Internet capacity transmission, computing processing, storage capacity and cryptography security, have fostered a technological and logical leap from the previous currency standards.

We refer to cryptocurrencies or crypto assets as the omni comprehensive family of digital tokens, which can be separated into *native coins* and *crypto tokens*. Native coins, like Bitcoin, represent a new asset class of electronic money universally accessible via peer-to-peer payment networks. Instead, crypto tokens are forms of "digital vouchers" that allow the token holders to get access to almost any type of service and assets: from monetary rewards, or commodities to loyalty points to even other cryptocurrencies. A token can either be fungible or non-fungible. Probably, at the moment, the most famous example of non-fungible tokens that hit the headlines are CryptoKitties. Each CryptoKitty is represented in the form of a non-fungible ERC-721 token, which allows for each entity to have specific attributes ("phenotype") determined by its immutable genes ("genotype") stored in the Ethereum smart contract.

The creation of new tokens is generally a less complex process than creating native coins as it does not require to modify the codes from a particular protocol or create a new blockchain from scratch. Moreover, the recent implementation of blockchain middleware and app development tools, Turing-complete codes for smart contracts on the blockchain allow crypto tokens to be easily created, published, shared and exchanged.

This chapter will not provide a taxonomy of cryptocurrencies but it will rather focus on crypto tokens as alternative funding instruments of new token-based business models. For a taxonomy of cryptocurrencies, we refer the readers to Bech and Garratt (2017).

9.2   Native Digital Assets

Native tokens are digital fungible assets created within a novel or "forked" off a pre-existing blockchain. A native token "a" exists and operates on the blockchain network "A" which allows peer-to-peer (sometimes, anonymous or pseudo-anonymous) transactions of "a" between different network participants. However, the reader should be aware that a native token needs a blockchain but a blockchain can function even without a token (Tasca and Tessone 2018).

In order to present the main features of native coins, we take the configuration proposed by Tasca (2016) who highlights a dual nature of Bitcoin: as a currency and as a payment network. In its first nature, Bitcoin operates as a currency. According to economic theory, a currency has three main features: it is a medium of exchange, a unit of account and a store of value. These frameworks can be extended beyond Bitcoin to analyse the characteristics of any other native coin.

Whether native coins could be considered currencies or not is an ongoing debate. As a matter of fact, the European Central Bank (ECB) and other financial market authorities do not confirm this view. For this reason, these institutions do prefer to use the term "digital tokens" instead of "cryptocurrencies" when referring to native coins. This thesis is supported by the fact that digital assets do not ensue a legal tender (Tasca 2015). Within a given jurisdiction, a legal tender is mandatory accepted, accepted at full-time value and it has the power to release debtors from paying their obligation. However, some jurisdictions have already begun the process towards legalisation of cryptocurrencies. For instance, in Japan the Financial Services Agency is working towards the full regularisation of cryptocurrencies as a legal means of payment (Terazono 2017). Switzerland is also very advanced in this respect (FINMA 2018).

Despite the fact that native coins are not always perfectly designed and implemented, it is undoubted that they bring some features typical of money. Back to Bitcoin again, we can say that it acts as a means of exchange and allows counterparties to avoid the "coincidence of the wants"[1]: the number of daily transactions has grown over time from around 1000 in 2011 to around 200,000 in 2018. At the same time, Bitcoin is a unit of account since it is divisible (the smallest possible unit is called Satoshi: 1 satoshi = 0.00000001 Bitcoins), fungible and countable. At the same time, Bitcoin's deflationary property prevents it from being considered as a good store of value. The number of Bitcoins issued over time is destined to decrease geometrically with 50% reduction every 4 years (Tasca 2015). That being said, from a pure monetary viewpoint, native coins do not generally fulfil the properties of money (Tasca 2016).

A novelty of native coins with respect to more traditional forms of money is that they come together with a network infrastructure that enables a disintermediated peer-to-peer exchange of coins. They combine together the characteristic of money with those of the payment systems. To better understand this aspect, the framework proposed by Bradford and Keeton (2012) can be taken into consideration. It identifies four main relevant features of a payment network: speed, payer control, security and universality. As a matter of fact, a Bitcoin transaction takes 1 + hours to be settled in the ledger. However, other coins are much faster. Ripple, for example, takes 4 seconds per transaction to be registered (Morgan 2018). With respect to payer control feature, there is no limit as cryptocurrencies can easily and quickly be transferred without any working hours constraints from wallet app or other operators. From a security perspective, transfers in cryptocurrency networks happen

---

[1] Coincidence of wants (also known as "double coincidence of wants") occurs when the supplier of good A is a demander of good B and vice versa. Without a medium of exchange, trades would be limited to this situation only (Jevons 1876; Ostroy and Starr 1990).

between hashed addresses so the risk of unauthorised transactions is very low (Antonopoulos 2014). However, wrong transactions cannot be cancelled but only adjusted with other transactions of opposite sign. Finally, from the perspective of universality, although cryptocurrencies count on a smaller network when compared with more traditional payment systems, we need to highlight the constantly growing trend of users that opt for cryptocurrency payment systems. It has been roughly estimated that, as of March of 2017, the number of active users of Bitcoin wallets was in the range of 2.9 million and 5.8 million (Hileman and Rauchs 2017). However, since then, proportional to the market valuation and price of Bitcoin, the cryptocurrency's user base has grown at a rapid rate. Coinbase alone, the global market's largest bitcoin brokerage and wallet platform, serves more than 13 million active users.

That being said, one should also consider that cryptocurrency payment networks are stand-alone systems: each native coin functions within its unique payment network without any possibility to interact with other networks. For this reason, the interoperability between different blockchain systems remains one of the major future challenges to be addressed (Bridgwater 2018). In this respect, it is worth mentioning that some new technological solutions have been proposed to overcome this problem, see, for example, Sidechains (Back et al. 2014) and Quant Overledger (Verdian et al. 2018).

9.3   Crypto Tokens

Since 2008, when an inventor (or group of inventors) under the pseudonym of Satoshi Nakamoto introduced Bitcoin (Nakamoto 2008), many other cryptocurrencies have been introduced by leveraging on the original Nakamoto's protocol or by elaborating new ones. The recent technological improvements have enhanced the number of applications of blockchain through smart contracts to automatically move digital assets according to arbitrary pre-specified rules (Buterin 2014). Specifically, crypto tokens give the opportunity to create businesses and automate them while maintaining the record of the different states of data exchanged in the blockchain. Token Market provides a quite exhaustive list of the tokens available in the market.

A commonly accepted taxonomy—adopted by many institutions including the Swiss Financial Market Supervisory Authority (FINMA 2018)—dentifies three main token classes:

- Payment tokens: these are synonymous with cryptocurrencies, intended as a means of payment for acquiring goods or services or as a means of money or value transfer;
- Utility tokens: these are intended to be the only way to provide digital access to applications and/or services (generally) built on the top of blockchain-based infrastructures.
- Asset/Debt tokens: they have a similar role as a share (Tasca et al. 2018), and for the investor they represent assets such as a debt or equity security owned.

There is then a fourth type of tokens (i.e., Hybrid) which are characterised by a mixture of the previous three features.

This classification does not implement a rigid distinction between native coins and tokens but it classifies important tokens, which will be specified later in the chapter.

Tokens play a vital role in the crowdfunding process of platform-based businesses and have been recently adopted by startups seeking to bypass the complicated and costly auditing and regulatory burden surrounding traditional funding models via banks or venture capitalists. Tokens represent then a means to raise funds from both platform users and sophisticated investors (Tasca et al. 2018). Much simpler than an Initial Public Offering (IPO), the Initial Coin Offering (ICO) process is composed of three distinctive phases:

1. The white paper announcement: the initial report or proposal where the company presents to potential investors and supporters of the business and other important features.
2. The release of tokens: often issued via smart contract whose code is public. Usually, the token generation is composed by two subphases: pre-allocation, granting a discount on the purchase of tokens and allocation, at full price.
3. Token listing: Complete the ICO, tokens are listed in one or more exchanges.

9.4   TOKEN-BASED BUSINESS MODELS

Having an idea is useless if one does not have enough capital to translate it into a reality. That is why capital raising is a vital process for any entrepreneurial endeavour, which allows the entrepreneurs to get the business off the ground or help them in the daily operations or business development.

With regards to capital raising, the last number of years have witnessed an exponential adoption of alternative token-based funding models. The lack of regulation and the relatively easy process of token creation engender the perfect conditions for a new funding trend: companies, especially start-ups, instead of raising funds through the traditional channels (equity issuing or taking out a loan) have been selling tokens in the market to the public and bootstrapping their own project based on the proceeds collected with the token allocations. There has been a massive adoption of this solution resulting in the proliferation of token-based business models. At the moment of writing we count about 800 tokens, which means an equivalent number of token-based business projects worldwide. Just to name a few, Nexo (www.nexo.io) is a token-based business that offers the opportunity to provide crypto-backed loans. Another example is Augur (www.augur.net) which is a decentralised oracle and prediction market. Coinlion (coinlion.com) distributes tokens to those who share information related to the portfolio management and trading of cryptocurrencies.

Apart from a few notable projects, the quality of token-based business models, whose number has skyrocketed in the last number of years, is generally very low (Tasca and Widmann 2017). Moreover, scams and frauds occur regularly. According to a recent study, 25% of the projects default in about 50 days after their token being listed in public trading markets (Tasca et al. 2018). As reported by Fortune, nearly half of ICOs started in 2017 failed by February 2018 (Morris 2018).

In order to protect investors and limit these frauds and excessive risks, regulators have started to develop the first regulatory frameworks and to promulgate the first official laws (Clayton 2017). On the other hand, a positive aspect of the token-based funding models is that investors are not locked-in for months or years as in the traditional VC market. Instead, tokens are tradable in the secondary market generally after a few weeks from the date of the ICO.

9.5   Driving Forces Behind the Token-Based Business Models

In the previous section, we have seen that new business models are designed and built around the concept, the meaning and the utility of "brand coins", which represent alternative funding instruments for the platform economy (Hayes and Tasca 2016). This is a remarkable novelty that stems from four major trends: (1) platform business models, (2) peer-to-peer networks, (3) open-innovation, and (4) crowdfunding.

Platform business models: These are "intermediaries that connect two or more distinct groups of users and enable their direct interaction" (Zhu and Furr 2016, p. 4). More recently, these platforms deal in not only market-mediated digitally-encoded information such as software, music and banking services, but also goods and services more generally. While, the first-generation platforms were online or digital only, the second generation of platforms has emerged operating "online to offline" (O2O) throughout the economy. Uber, Airbnb and Caviar are just a few examples of the myriad of O2O platforms operating across different sectors.

Three elements are recognised to make a platform business model successful (Boncheck and Choudary 2013):

- The *Toolbox*. It creates a connection by making it easy for others to plug into the platform;
- The *Magnet*. It creates a pull that attracts participants to the platform. For transaction platforms, both producers and consumers must be present to achieve critical mass;
- The *Matchmaker*. It facilitates the connections between producers and consumers or lenders and borrowers.

Most successful internet-based businesses recently developed, have adopted the platform business model because they use technology to connect people, organisations and resources in an interactive ecosystem in which value can be created and exchanged. In these cases, the companies scale up by building on their networks of users instead of accumulating inventories (Parker et al. 2016).

Peer-to-peer (P2P) networks: A peer-to-peer network is "group of computers, each of which acts as a node for sharing files within the group" (Technopedia). This form of network blossomed during the progressive and constant increase in bandwidth Internet capacity registered during the 80s, 90s and 2000s (Oram 2001). P2P networks are built on disintermediation and share of content. Indeed, the robustness of the network itself is not provided by a single central entity or restricted group of peers anymore.

Open-innovation: This is important because it highlights organisations' need for a more enlightened role for R&D in a world of abundant information, better managing and accessing intellectual property, increasing future business (Chesbrough 2006).

Crowdfunding: Defined as an open call over the Internet for financial resources in the form of a monetary donation, sometimes in exchange for a future product, service or reward (Kleemann et al. 2008). The slow action by national and international regulators left a wide legislative space to new venture attracted by the opportunity to collect easily significant amount of funds from the retail market in non-traditional and not monitored ways. There are many facets of crowdfunding: (1) lending-based crowdfunding, which consists of loans which are repaid with interest, (2) equity-based crowdfunding in which investors receive shares of the startup company, (3) reward-based crowdfunding that involves rewarding funders with a product that has actual monetary value, often an early version of the product or service being funded, and (4) donation-based crowdfunding in which backers donate funds because they believe in the cause (Pelizzon et al. 2016). See Chapter 1 for further discussion of crowdfunding.

Token-based business benefits from those four drivers as they build on the principles of the platform models by providing a digital means of exchanging information and value. At the same time, they leverage the latest innovations in the blockchain space that enhance the potential of P2P decentralised networks. Moreover, in line with the open-innovation trend described above, blockchain software is generally based on an open-source license model. The open-source model allows everyone to audit and improve the source code of protocols and smart contracts. For example, Ethereum is licensed under GNU LGPLv3, Bitcoin Core is licensed under the MIT License and Hyperledger Fabric is licensed under Apache 2.0. At the same time, blockchain-based systems leverage a crowdsourced means of verifying transactions. In that sense, it is like Wikipedia, where community consensus governs what information is trusted to be accurate. This aspect of token-based business models refers not so much to open source, but to open culture based on the participation of the crowd brought together to build, exchange and share.

9.6   CRYPTO TOKENS TO ENHANCE THE SHARING ECONOMY

There is no official definition of the Sharing Economy because it is a concept adapted to the context of reference. Some proxies enclose ideas of the Sharing Economy. Take for instance "access-based consumption" which represents a set of "transactions that can be market mediated but where no transfer of ownership takes place and differ from both ownership and sharing" (Eckhardt and Bardhi 2015). Also "collaborative consumption" is a proxy for Sharing Economy. In this respect, it can be defined as a "peer-to-peer-based activity of obtaining, giving, or sharing the access to goods and services, coordinated through community-based online services" (Hamari et al. 2016, p. 2049).

Regardless of the different conceptual forms, it is inevitable that the Sharing Economy challenges the current economic institutional framework by shifting from a framework that protects people from each other, to a framework that helps people trust each other via the "trust machine" (*The Economist* 2015). This is achieved thanks to universally acceptable censorship-resistant ledgers that solve the double spend problem and allow anonymous users to achieve consensus on the states of the shared ledgers.

The blockchain platforms that empower the token-based business models have the potential to lead us towards a new economic paradigm: a shift from centralised to decentralised online marketplace solutions and from centralised two-sided platform business models—which market-mediate suppliers (lenders) and consumers (borrowers)—to P2P blockchain-based platforms on the top of open and decentralised networks where users are also producers/shareholders and where the value created is fairly and transparently redistributed.

We are already living in the so-called economy of Collaborative Commons characterised by the prevalence of sharing over ownership. This major structural and cultural change mainly applies to fungible products and services that can be easily standardised and automated, similar to the broad spectrum of services offered by traditional banks. But this is not necessarily the case. These platforms allow also for non-fungible assets and services to be exchanged. For example, Rent and

Runway is a platform that enables women to rent unique clothing and personal accessories online.

As this new economic paradigm is unfolding, now a hybrid economy is flourishing where some industries based on the Commons are starting to operate at near zero marginal cost, while other industries continue to cling to capitalist consumer markets. Companies like Uber and Airbnb will attempt to bridge the gap between the two economies and take advantage of both. Though, it is very likely that over the coming few years, new business models based on decentralised "dumb" platforms, such as Citi Arcade and LaZooz will continue to disrupt the Uber-like "smart" platform business models. Blockchain allows buyers and sellers or lenders and borrowers to do business directly with each other, without the intervention of a large commercial platform. In other words, as Ethereum founder Vitalik Buterin has put it: Blockchain does not make the taxi driver lose his job, the network technology makes Uber superfluous.

The introduction of token networks based on blockchain technology in such a dynamic environment could represent an additional stage towards a completely disintermediated sharing economy and distributed business models where the lines between users, producers and investors are blurred. Decentralied Collaborative Commons will expand across lateral networks, and as access will overcome ownership, competition will be superseded by cooperation, buyers (borrowers) and sellers (investors) will transition to *prosumers*.

Utility tokens will play an active role in this new system. A consumer who buys a utility token supports the network stability and liquidity. The more purchases and sales of services or goods happen in the network, the more effective the network will be. The use of utility tokens by new users increases the value of the tokens and consequently the investment value of the other users. More importantly, an investor using the utility token will increase the value of its investments while providing a better network for another user. Therefore, the distinction between stakeholders will fade: a customer will be an investor and vice versa. A business company based on utility tokens will potentially be favoured by a positive escalation effect where use of tokens will benefit users and platform originating a self-enforcement mechanism.

To conclude, we also want to emphasise that a token-based economy is not immune from hazards. Cascade-effects, scams and hoarding movements remain main risks to be addressed in order to smoothly evolve towards an authentic Sharing Economy.

References

Antonopoulos, A. M. (2014). *Mastering Bitcoin: Unlocking digital cryptocurrencies*. O'Reilly Media, Inc.

Back, A., Corallo, M., & Dashjr, L. (2014). *Enabling blockchain innovations with pegged sidechains*. Available at: https://blockstream.com/sidechains.pdf. Last accessed 16 August 2018.

Bech, M. L., & Garratt, R. (2017). Central Bank Cryptocurrencies. *BIS Quarterly Review*. Available at SSRN: https://ssrn.com/abstract=3041906.

Bonchek, M., & Choudary, S. P. (2013). Three elements of a successful platform strategy. *Harvard Business Review, 92*(1–2).

Bradford, T., & Keeton, W. R. (2012). *New person-to-person payment methods: Have checks met their match?* Economic Review-Federal Reserve Bank of Kansas City, 41.

Bridgwater, A. (2018). Blockchains are verticalizing, so we need interoperability. *Forbes*. Available at: https://www.forbes.com/sites/adrianbridgwater/2018/02/07/blockchains-are-verticalizing-so-we-need-interoperability/#4a7600417ab9. Last accessed 16 August 2018.

Buterin, V. (2014). A next-generation smart contract and decentralized application platform. *White Paper*. Available at: https://cryptorating.eu/whitepapers/Ethereum/Ethereum_white_paper.pdf. Last accessed 16 August 2018.

Chesbrough, H. W. (2006). *Open innovation: The new imperative for creating and profiting from technology*. Harvard Business Press.

Clayton, J. (2017). *Statement on cryptocurrencies and initial coin offerings*. World.

Eckhardt, G. M., & Bardhi, F. (2015). The sharing economy isn't about sharing at all. *Harvard Business Review, 28*(1).

FINMA. (2018). *FINMA publishes ICO guidelines*. Available at: https://www.finma.ch/en/news/2018/02/20180216-mm-ico-wegleitung/. Last accessed 16 August 2018.

Hamari, J., Sjöklint, M., & Ukkonen, A. (2016). The sharing economy: Why people participate in collaborative consumption. *Journal of the Association for Information Science and Technology, 67*(9), 2047–2059.

Hayes, A., & Tasca, P. (2016). Blockchain and crypto-currencies. In S. Chishti and J. Barberis (Eds.), *The FinTech book*. Chichester: Wiley.

Hileman, G., & Rauchs, M. (2017). *Global blockchain benchmarking study*. Cambridge Centre for Alternative Finance, University of Cambridge. Available at SSRN: https://ssrn.com/abstract=3040224.

Jevons, W. S. (1876). *Money and the mechanism of exchange*. New York: D. Appleton & Co.

Kleemann, F., Voß, G. G., & Rieder, K. (2008). Un(der) paid innovators: The commercial utilization of consumer work through crowdsourcing. *Science, Technology & Innovation Studies, 4*(1), 5–26.

Menger, K. (1892). On the origin of money. *The Economic Journal, 2*(6), 239–255.

Morgan, J. P. (2018). *Decrypting cryptocurrencies: Technology, applications and challenges.* Available at: https://forum.gipsyteam.ru/index.php?act=attach&type=post&id=566108. Last accessed 16 August 2018.

Morris, D. (2018). Nearly half of 2017's cryptocurrency 'ICO' projects have already died. *Fortune.* Available at: http://fortune.com/2018/02/25/cryptocurrency-ico-collapse/. Last accessed 16 August 2018.

Nakamoto, S. (2008). *Bitcoin: A peer-to-peer electronic cash system.* Available at: https://bitcoin.org/bitcoin.pdf. Last accessed 16 August 2018.

Oram, A. (2001). *Peer-to-Peer: Harnessing the power of disruptive technologies.* O'Reilly Media, Inc.

Ostroy, J. M., & Starr, R. M. (1990). The transactions role of money. In *Handbook of monetary economics* (Vol. 1, pp. 3–62).

Parker, G. G., Van Alstyne, M. W., & Choudary, S. P. (2016). *Platform revolution: How networked markets are transforming the economy and how to make them work for you.* W. W. Norton & Company.

Pelizzon, L., Riedel, M., & Tasca, P. (2016). Classification of crowdfunding in the financial system. *New Economic Windows,* pp. 5–16.

Samuelson, P. A. (1958). An exact consumption-loan model of interest with or without the social contrivance of money. *Journal of Political Economy, 66*(6), 467–482.

Tasca, P. (2015). Digital currencies: Principles, trends, opportunities, and risks. *SSRN Electronic Journal,* 1–110. https://doi.org/10.2139/ssrn.2657598.

Tasca, P. (2016). The dual nature of Bitcoin as payment network and money. *VI Chapter SUERF Conference Proceedings.* Available at SSRN: https://ssrn.com/abstract=2805003.

Tasca, P., & Tessone, C. J. (2018). *Taxonomy of blockchain technologies.* Principles of identification and classification. Available at SSRN: https://ssrn.com/abstract=2977811.

Tasca, P., & Widmann, S. (2017). The challenges faced by blockchain technologies—Part 1. *Journal of Digital Banking, 2*(2), 132–147.

Tasca, P., Vigliotti, M. G., & Gong, H. (2018). Risks and challenges of initial coin offerings. *Journal of Digital Banking, 3*(1), 81–96.

Terazono, E. (2017). Bitcoin gets official blessing in Japan. *Financial Times.* Available at: https://www.ft.com/content/b8360e86-aceb-11e7-aab9-abaa44b1e130. Last accessed 16 August 2018.

The Economist. (2015). The trust machine. *The Economist.* Available at: https://www.economist.com/leaders/2015/10/31/the-trust-machine. Last accessed 16 August 2018.

Verdian, G., Tasca, P., Paterson, C., & Mondelli, G. (2018). Quant overledger whitepaper. Available at: https://objects-us-west-1.dream.io/files.quant.network/Quant_Overledger_Whitepaper_v0.1.pdf. Last accessed 16 August 2018.

Zhu, F., & Furr, N. (2016). Products to platforms: Making the leap. *Harvard Business Review, 94*(4), 72–78.

CHAPTER 10

Blockchain Beyond Cryptocurrencies

Pierangelo Rosati and Tilen Čuk

Abstract It is claimed that blockchain technology has the potential to revolutionise how financial services firms conduct their business. This chapter presents the main characteristics of blockchain technology and summarises the extant research around the potential implications of blockchain adoption for four main financial activities: payments and remittance, credit and lending, trading and settlement, and compliance. Current gaps in the literature are discussed in order to identify avenues for future research.

Keywords Blockchain · Distributed ledger technology

P. Rosati (✉)
DCU Business School, Dublin City University, Dublin, Ireland
e-mail: pierangelo.rosati@dcu.ie

T. Čuk
Centre Perelman de Philosophie du Droit,
Université Libre de Bruxelles, Brussels, Belgium

© The Author(s) 2019   149
T. Lynn et al. (eds.), *Disrupting Finance*, Palgrave Studies
in Digital Business & Enabling Technologies,
https://doi.org/10.1007/978-3-030-02330-0_10

10.1 Introduction

The financial services industry plays a key role for businesses and society since it enables saving and investment, provides protection from risks and supports the creation of new jobs and enterprises (World Economic Forum 2013, 2016). Developments in information technologies have changed the industry over time by enabling an enormous increase in transactions and diversified products (Gardner 2011). However, the pace of innovation in the sector has traditionally been very slow (Gardner 2011; Michel 2014). This is mostly due to the regulatory burden and to the conservative culture embedded within the industry (Gardner 2011; Michel 2014; Das et al. 2018). The result is a linear and predictable innovation pattern (Luftenegger et al. 2010) with only five major technological innovations in a 50-year period, namely: (1) computerised information systems in 1950s (Luftenegger et al. 2010), (2) automatic teller machines (ATMs) in 1960s (Batiz-Lazo 2009), (3) electronic stock trading in 1970s (Terrell 2010), (4) mainframe computers in 1980s, and (5) the Internet in the 1990s/early 2000s (Desai 2015).

However, things have changed significantly over the last decade. As the industry moves to what the International Data Corporation refers to as the "third platform"—a technology trend towards ubiquitous computing, big data, and the widespread adoption of social and mobile technologies (IDC 2012) in response to customer expectations for more innovative and personalised products. Regulatory changes such as, for example, the new EU Payment Service Directive (PSD2), which is discussed in Chapter 7 of this book, aim to significantly lower the barriers to market entry, therefore increasing competition. These recent changes have triggered what scholars and practitioners refer to as the "FinTech revolution" (Mackenzie 2015; Gomber et al. 2018), built around three main pillars: (1) capital availability both for start-ups in the form of venture capital and for incumbents; (2) new technologies; and (3) new business models (Gomber et al. 2018).

As an enabling and disruptive technology, blockchain is arguably at the core of the FinTech revolution and has the potential to radically change a large number of activities and processes within the industry. These changes are expected to provide huge improvements in efficiency, generating potential savings of $16–20 billion a year (Santander 2015; Capgemini 2016).

Blockchain is mostly known as the technology underpinning Bitcoin, "*a payment system based on cryptographic proof instead of trust*" (Nakamoto 2008, p. 1), also referred to as cryptocurrency. While cryptocurrencies have been discussed in Chapter 9, this part of the book explores other blockchain applications for the financial sector and discusses related literature gathered from the finance, information systems and computer science fields. The rest of the chapter is structured as follows. The next section provides an overview of blockchain technology. This is followed by a discussion of the current challenges that the technology poses for financial services firms, and the impact of blockchain on four main financial activities: (1) payments and remittance, (2) credit and lending, (3) trading and settlement, and (4) compliance. The chapter concludes with a discussion of further avenues for research.

10.2   What Is Blockchain?

Commercial transactions have been stored in ledgers since ancient times. Initially kept on clay tablets or papyrus, they then moved to paper and ultimately to bytes with the advent of computerised information systems (Rosati and Paulsson 2017). Regardless the format in which ledgers were kept, they have traditionally relied on human inputs; as such, ledgers have been prone to errors which translate into additional costs and inefficiencies for organisations and for the economic system as a whole. Digital distributed ledgers promise to fix these issues through a unique combination of distributed networks and cryptography. Blockchain technology is by far the most famous example of distributed ledgers technologies (Beck et al. 2017). It was first launched in 2008 by either a programmer or a group of programmers under the pseudonym of Satoshi Nakamoto (Nakamoto 2008) and today is mostly known for being the technology underpinning Bitcoin, arguably the most famous open-source peer-to-peer digital currency (Mougayar 2016). However, potential applications of blockchain extend beyond digital currencies, potentially impacting corporate governance, social institutions, democratic participation, and the functioning of capital markets (Wright and De Filippi 2015).

Blockchain can be defined as a decentralised, transactional database that enables validated, tamper-resistant transactions across a large number of participants (i.e. nodes) in a network (Glaser 2017;

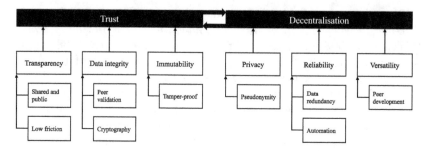

Fig. 10.1 Characteristics of blockchain (Adapted from Seebacher and Schüritz [2017])

Beck et al. 2018). But blockchain is not just a technological innovation. By providing a valid alternative to traditional trusted intermediaries, it also carries philosophical, cultural, and ideological underpinnings that have to be taken into account (Mougayar 2016). For this reason, Mougayar (2016) proposes to integrate the technical definition with a business definition and a legal definition. From a business point of view, blockchain can be defined as a peer-to-peer exchange network for transferring value, while, from a legal perspective, it can be defined as a technology to validate transactions.

Blockchain has two main characteristics (Fig. 10.1). First, it brings trust where there is none (Beck et al. 2016; Rosati et al. 2016; Tapscott and Tapscott 2016; Glaser 2017). In fact, blockchain-based systems ensure higher transparency by making information available to all network participants, but they also leverage cryptography and peer validation of transactions to ensure data integrity and record immutability (Böhme et al. 2015; Sun et al. 2016). Second, blockchain-based systems are fully distributed. Users' privacy is protected by using pseudonyms while the system's reliability is ensured by storing a copy of the database in every node (Böhme et al. 2015; Beck et al. 2016; Weber et al. 2016). These two key characteristics of blockchain are ultimately interconnected as mechanisms to build trust are needed for creating a decentralised network, and decentralisation provides the means for users to get involved in the network by establishing the basis for a consensus mechanism (Seebacher and Schüritz 2017).

Users' ability to read and submit transactions to a blockchain depends on their access to transactions. There are different "flavours" of

Table 10.1   Types of blockchain (Adapted from Beck et al. [2018])

	Access to transaction validation	
Access to transactions	*Permissioned*	*Permissionless*
Public	All nodes can read and submit transactions. Only authorised nodes can validate transactions—e.g. Ripple	All nodes can read, submit and validate transactions—e.g. Bitcoin
Private	Only authorised nodes can read, submit and validate transactions—e.g. R3 Corda	Not applicable

blockchain (Table 10.1). In a public (permissionless) blockchain, every node can read and submit (validate) transactions while in a private (permissioned) blockchain only authorised nodes can (Peters and Panayi 2016).

Regardless of the type of blockchain, there are some key features that are common (although to different extents) to all of them (Mougayar 2016):

- Distributed network: the adoption of a blockchain removes all centralised entities and distributes access to all of the participants (i.e. nodes) in the network. In other words, all participants in the network, and not a specific one, can verify the transactions. Miners are key actors in this distributed network as they work to solve the computational problems that allow to create, verify, and securely store transactions.
- Cryptography: it enables parties to maintain the privacy of the information they send to each other. Blockchain uses Public Key Infrastructure (PKI) mechanisms to execute transactions. Each blockchain user has a public key and a private key. In order to complete a transaction, a sender needs to know the public key of the receiver who can decrypt the message by using its own private key. Every transaction is stored in a block, which has a unique fingerprint (i.e. hash) that ensures the authentication of the transaction source.
- Timestamp: every transaction that occurs on the blockchain is timestamped and no one is able to change it once it has been recorded.

Blockchain is mostly known for its ability to process monetary and financial transactions. However, it can also ensure that transactions comply with specific rules parties have agreed upon in the form of "smart contracts" (Tschorsch and Scheuermann 2016; Risius and Spohrer 2017). Beck and Müller-Bloch (2017) refer to blockchains supporting this kind of applications as "blockchain 2.0".

Despite all the hype around blockchain, it is still a nascent technology and there are technical, non-technical, and regulatory challenges to overcome in order to foster adoption. Extant academic research has focused on technical aspects of blockchain (Yli-Huumo et al. 2016; Risius and Spohrer 2017) but a number of issues, such as efficiency, latency, throughput, scalability, security, and systems integration still need to be (partially) addressed (Yli-Huumo et al. 2016). Security issues represent real concerns for financial institutions as they store and exchange highly sensitive information about their customers and operate under strict regulations. In addition, the new European General Data Protection Regulation[1] (GDPR) now requires organisations to obtain specific consent from their clients to use their private information. With permissionless public blockchains, it is difficult to control who accesses the blockchain. As such, financial institutions are more likely to adopt a private or permissioned blockchain than a public blockchain (Fink 2017). However, due to smaller number of participants (i.e. nodes), private or permissioned blockchains are more vulnerable to 51-percent attacks[2] (Peters and Panayi 2016; Yli-Huumo et al. 2016). Further research is arguably needed in this space in order to increase technology adoption in the financial sector.

Non-technical challenges, instead, are mostly related to (a) building up innovation legitimacy (Lynn et al. 2018), (b) understanding the determinants of users' adoption, (c) measuring the value generated by blockchain investments, and (d) assessing potential impacts on society (Risius and Spohrer 2017). Finally, regulatory challenges mostly arise from the distributed nature of blockchain applications which, by definition, can span across multiple jurisdictions, with responsibilities for system maintenance are shared among all network participants (Yeoh 2017).

---

[1] See https://gdpr-info.eu/ for further details.

[2] "The ability of someone controlling a majority of network hash rate to revise transaction history and prevent new transactions from confirming" (Bitcoin.org 2018).

10.3   Payments and Remittance

Departing from cryptocurrencies, here we mainly focus on interbank and cross-border payments which are often processed by intermediary clearing firms. These processes require a series of complicated processes including bookkeeping and transactions and balance reconciliations across multiple financial institutions[3] (Guo and Liang 2016). The result is a long and time-consuming process which often translates in delays in payments settlement and additional costs. Cross-border payments totalled $27.7 trillion in the first quarter of 2017 (BIS 2017), and represent 20% of total transaction volumes in the payments industry and 50% of the revenues (McKinsey 2016). Notwithstanding the scale, 43% of the capital transferred is lost in transaction costs (Guo and Liang 2016).

Beside the actual transaction cost, the time delay between payment initiation and settlement creates a risk for the parties involved; this is mostly related to the risk default of the counterparty and to fluctuations of the foreign currency whose value is determined by market rules (Bott and Milkau 2017). By enabling peer-to-peer payments and by offering 24/7 settlements, blockchain can reduce transaction costs and risk while bringing (almost) real-time settlements and increased transparency and traceability (Buitenhek 2016). Given these undeniable benefits, it is not surprising that both central banks and private institutions have started looking at blockchain-based applications for payments (Bott and Milkau 2017).

Inefficient payments processing is a business as much as an ethical problem. In fact, cross-border multi-currency payments are not only associated with business transactions; remittance also accounts for a significant amount of cross-border money transfers. According to recent World Bank estimates, international remittance totalled $585 billion in 2017, 7.32% of which was lost in transfer fees (World Bank 2017). In addition, 39% of the world's population, mostly comprised of the population in developing countries, does not have a bank account making it very hard for the receiver to actually collect the money being transferred (Mesropyan 2016). In this context, thoughtful combinations of blockchain systems and mobile technologies could potentially put billions back

---

[3] See, for example, Park (2016) for more details on the current cross-border payment settlement process.

into the pockets of families in developing countries therefore reducing (at least partially) the gap with richer countries.

10.4   CREDIT AND LENDING

Credit and lending is another area where blockchain can significantly change current operations. Other chapters in this book discuss peer-to-peer lending (Chapter 2) and crowdfunding (Chapter 1), two well-established innovations in the FinTech landscape with significant growth rates. Despite the attention they receive today, these practices are essentially as old as commerce; the main difference is that they were traditionally based on informal and interpersonal trust relationships. Peer-to-peer platforms have essentially found a way to reduce information asymmetry and formalise the relationship between parties therefore increasing trust between the two sides of the market, and, of course, to profit from their intermediation role. As a technology enabling peer-to-peer "trust-free" transactions, blockchain can replace both traditional (e.g. banks, credit unions) and new intermediaries (e.g. peer-to-peer lending platforms) therefore lowering transaction costs of lending and business financing (Larios-Hernández 2017). A typical example is the adoption of blockchain-based tokens to enable disintermediated crowdfunding campaigns also known as Initial Coin Offerings[4] (ICOs) (see, among others, Rohr and Wright 2017; Adhami et al. 2018; Catalini and Gans 2018; Chen 2018; Howell et al. 2018). Despite all the attention these blockchain-enabled peer-to-peer systems are receiving from investors, regulators, and the media, the volume of capital that goes through these channels still represents a tiny portion of the overall credit/lending market. As Hawlitschek et al. (2018) suggest, these systems still struggle in crossing users' "trust frontier". The authors further suggest that a widespread adoption of these systems depends on the development of trusted interfaces.

In the context of credit and lending, blockchain could also be leveraged to improve lenders' decision-making. Risk assessment of potential borrowers (be it a company or an individual) is usually based on the historical records of financial transactions. Data availability and quality, however, pose significant challenges to the validity and robustness of

---

[4]This topic is discussed in more details in the previous chapter.

credit score models (Abdou and Pointon 2011). These issues are particularly pronounced when potential borrowers are small- and medium-sized businesses or individuals for which information is rarely publicly and/or readily available (Thomas et al. 2017). This ultimately results in inefficient capital allocation (Jacobson and Roszbach 2003) and lost growth opportunities (Beck and Demirguc-Kunt 2006).

In order to overcome these limitations, financial institutions have started looking at using "alternative data" in their models such as mobile-phone information, psychometric testing, social media activity, or ecommerce transactions (McEvoy 2014). However, collecting, aggregating, and integrating data from different sources can be challenging and require specialist skills that financial companies do not always have in-house. The growing demand for this kind of service has led to the emergence of data marketplaces (Stahl et al. 2014). Data marketplaces usually offer a wide range of capabilities, such as data gathering, aggregation, integration, processing, enrichment, etc. (Roman and Gatti 2016). When it comes to credit scoring, data marketplaces can represent a common point of entry for perspective lenders and borrowers through which information is exchanged securely (Roman and Gatti 2016). Data marketplaces are essentially centralised systems, therefore require different stakeholders to trust a third-party managing their data. This may prove to be particularly challenging for very sensitive and potentially valuable data like those used for credit scoring. In this context, blockchain can be leveraged to create disintermediated trusted data marketplaces that securely connect together information providers, perspective borrowers and lenders, and guarantees data provenance and data integrity (Roman and Gatti 2016). Blockchain-enabled systems have therefore the potential to improve the credit scoring processes, therefore lowering default rates and providing undoubted economic benefits (Byströrm 2016).

10.5   TRADING AND SETTLEMENTS

Chapter 8 in this book is dedicated to recent advancements in trading technology. Even though trade execution time has been brought down to milliseconds, post-trade settlement is still a lengthy and redundant process that spans over multiple days. Two-day (T + 2) or three-day (T + 3) settlement is still the industry standard but more complex transactions like syndicated loans can take up to three weeks (Chiu and

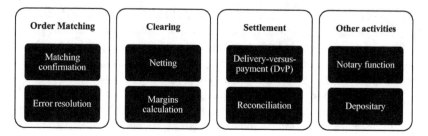

Fig. 10.2 Typical post-trade activities

Koeppl 2018). Figure 10.2 provides a brief summary of typical post-trade activities.[5]

As per payments, long settlement time and inefficient processes generate considerable costs and risk for the counterparties involved in a transaction. According to Broadridge (2015), industry spends $6 billion to $9 billion per year in core and ancillaries post-trade activities for standardised asset classes like equities and fixed income, but these figures go up to $24 billion when including more sophisticated asset classes and over-the-counter (OTC) markets.

Benos et al. (2017) argue that blockchain may impact the post-trade cycle in six ways:

- Reducing reconciliation and data management costs: the adoption of blockchain technology would allow the creation of a distributed, shared and synchronised database of security ownership. As such, it can simplify and automate most post-trade processes and significantly reduce the need for reconciliation. Mainelli and Milne (2016) estimate a potential 50% reduction for this kind of transaction costs.
- Flexible settlement times: intermediaries currently have at least a full day to prepare the settlement and borrow securities or cash as needed, therefore managing their own liquidity. A T + 0 (same-day) settlement, although desirable from a risk management perspective, would also require intermediaries to borrow cash or securities in advance, therefore increasing liquidity risk. In this context, flexible

[5] See AFME (2015) for more details.

settlement times appear more desirable in a blockchain environment, and could be implemented via smart contracts therefore creating benefits for all market participants. Khapko and Zoican (2018) demonstrate that a combination of flexible or short settlement cycles, coupled with option-like penalties for failures-to-deliver, would discipline competition on securities lending markets and improve market quality.

- Automated clearing: in a blockchain environment, when a trade is agreed the calculation of obligations (i.e. netting) could happen automatically and simultaneously therefore reducing the need for a clearing agent.
- Direct ownership: in the current market settings, investors are not always the owners of the securities they trade. There is indeed a chain of custodians who hold securities and act as intermediaries between issuers and investors. This creates implications for shareholder rights (Micheler 2015; Van der Elst and Lafarre 2018). When securities are issued in the form of (or can be transformed into) digital tokens, blockchain could facilitate direct ownership and increase transparency in the market, therefore enabling peer-to-peer trading.
- Traceability and transparency: blockchain is an "append-only" database. In other words, records cannot be deleted or altered once they have been stored in a block. This provides full traceability of transactions. The ledger is also shared among network participants, therefore increasing transparency. However, as Malinova and Park (2017) point out, investors often prefer privacy over transparency even when the latter may be socially desirable. Building on this point, the authors propose a blockchain-based market setting, which maximises social welfare, while protecting investors' privacy.
- Security and resilience: being a decentralised system, a blockchain does not have a single point of failure. As such, it is more resilient to cyberattacks and not subject to cybersecurity-related downtimes to the same degree.

Benos et al. (2017) also highlight a number of challenges to overcome before blockchain goes mainstream in the area of clearing and settlement. These are mostly related to (1) interaction between the digital and the physical world (e.g. current legacy assets held by custodians), (2) legal and regulatory limitations (e.g. proof of ownership), and (3) technology readiness (e.g. scalability and throughput). A number of initiatives have been undertaken to overcome regulatory barriers (Van der Elst and Laferre 2018). For example, the State of Delaware explicitly reference the use of blockchain technology in Section 224[6] of the general corporation law (DGCL) on July 21, 2017. Also, there are ongoing efforts from multiple stakeholders aiming to enhance blockchain performance and reliability (Yli-Huumo et al. 2016; Higgins 2018; Chiu and Koeppl 2018), which suggest that it will not be long before blockchain-based applications start moving from the proof-of-concept stage to the production stage. For example, the European Central Bank and the Central Bank of Japan have already conducted a first study to evaluate the possibility of using blockchain for real-time gross settlements that are crucial in conducting monetary policy (ECB 2017).

10.6   COMPLIANCE

Regulation is becoming increasingly burdensome in financial services. In order to increase investor protection and to prevent financial crime, post-crisis regulatory changes have dramatically increased the amount of reporting and compliance requirements for all the actors involved in the industry. Banks spent almost $100 billion for compliance in 2016 (McDowell 2017) and the overall expenditure is growing year-by-year (Thomson Reuters 2018). Regulatory technology (RegTech[7]) may be a way to reduce these costs and so financial services are investing a significant amount of resources in this direction (Spezzati 2017).

In this section, we will focus on three main regulatory frameworks: (1) the EU Markets in Financial Instruments Directive (MiFID II[8]) and the corresponding US Dodd–Frank Wall Street Reform and Consumer

---

[6] "Any records administered by or on behalf of the corporation in the regular course of its business, including its stock ledger, books of account, and minute books, may be kept on, or by means of, or be in the form of, any information storage device, method, or 1 or more electronic networks or databases (including 1 or more distributed electronic networks or databases) [...]" (Delaware General Corporation Law—Section 224).

[7] Chapter 6 covers this topic in more details.

[8] https://ec.europa.eu/info/business-economy-euro/banking-and-finance/financial-markets/securities-markets/investment-services-and-regulated-markets-markets-financial-instruments-directive-mifid_en.

Protection Act,[9] (2) Know Your Customer (KYC) and Anti-Money Laundering (AML) regulation, and (3) financial reporting standards (IFRS/IAS). Blockchain can help financial companies and regulators in handling compliance requirements across all these areas. MiFID II and the Dodd-Frank have been enacted in response to the global financial crisis with the main objective of increasing transparency in the financial markets and strengthening investor protection (Black 2010; Prorokowski 2015). Both regulations require financial firms to keep track of all interactions related to every single transaction. Comprehensive, traceable and time-stamped reporting is essential to comply with these regulations and this is where blockchain may represent a valuable solution. For example, Sheridan (2017) argues that a publicly available blockchain can be an effective solution to effectively communicate equivalence decisions under MiFID II; this would represent a single source of truth for identifying third-party countries that are allowed to conduct financial business in the EU. Similarly, a distributed ledger can be used by different regulators to uniformly record firm-by-firm authorisations and permissions.

Processes for KYC/AML compliance are particularly redundant. Financial institutions are required to onboard their customers before conducting any business activity with them in order to avoid working with/for customers involved in illegal activities (Ruce 2011). The onboarding process consists of an exchange of documents and information between a financial institution and the perspective customer. Even though most of the documents required for onboarding customers are standardised, the overall process has to be repeated by each institution for each customer with which it wants to interact. Secure and reliable information sharing could eliminate redundancies therefore making the process more efficient and improving customer experience (Moyano and Ross 2017). Moyano and Ross (2017) propose a system architecture for a distributed ledger through which financial institutions can verify the result of standardised KYC tasks that have already been conducted for a specific customer. Such a system would lower the cost associated with KYC processes without compromising participants' security and privacy.

Financial reporting quality is historically a key concern for regulators. Even though this applies to every industry, the financial sector has traditionally received special attention in this respect due to the key enabling

---

[9] https://www.govtrack.us/congress/bills/111/hr4173/text.

role that it plays in the economy. Standard setters are continuously trying to increase the transparency and accuracy of financial statements, and to unify financial reporting practices across multiple jurisdictions (Barth et al. 2008). There is still limited evidence of blockchain applications for financial reporting purposes (Dai and Vasarhelyi 2017) and some contrasting opinions in the literature (Coyne and McMickle 2017). However, some characteristics of the blockchain like data integrity, (almost) real-time updates, instant sharing of necessary information, and programmable and automatic controls may represent the basis for a new financial reporting ecosystem (Dai and Vasarhelyi 2017). Dai and Vasarhelyi (2017) present a first example of a blockchain-enabled triple-entry accounting information system, which may represent a step forward towards real continuous auditing. In a similar vein, Wang and Kogan (2018) propose a prototype of a blockchain-based transaction processing system for real-time accounting that leverages zero-knowledge proofs to find a trade-off between transparency and confidentiality. This kind of financial reporting systems could prove to be particularly suitable for the financial sector where every transaction has to be recorded and where traceability and records' immutability are extremely important to prevent fraud. Furthermore, auditors and regulators may be able to access financial records in real time if needed, therefore increasing transparency and timely interventions where needed.

10.7   CONCLUSION AND AVENUES FOR FUTURE RESEARCH

This chapter provides an overview of blockchain technology and of the extant literature on its potential applications and implications for the financial sector. Despite all the hype and all the promises around blockchain technology, it still remains an early-stage technology; the number of potential use cases is getting larger and larger but very few of them have made their way to the market. This is particularly the case for financial services where conservatism and regulatory requirements represent significant challenges for innovation.

Blockchain is, in essence, a technological innovation. Thus, perhaps unsurprisingly, most of the research so far comes from computer science domain. Even though pure technical aspects of blockchain go beyond the scope of this chapter, it is worth to briefly mention some existing technical challenges that also represent opportunities for research. First of all, integrating existing legacy systems with blockchain is still a very difficult task; it would be naïve to think that organisations will get rid of existing systems to move to blockchain-based systems. Therefore, integration/migration patterns have to be identified in order to streamline adoption. The trade-off between security, performance, and sustainability is another topic that is widely discussed. In relation to this, different combinations of block sizes and encryption and consensus mechanisms are being explored; different combinations are likely to be more suitable for some applications but not for others, hence the need for a contingency approach to this issue.

Although there are a number of question marks around blockchain as a technology, the number of technical studies is arguably growing fast (Yli-Huumo et al. 2016). The same cannot be said for organisational and business-related research (Risius and Spohrer 2017). In a conservative, heavily regulated and profit-driven sector like financial services, reducing the uncertainty around implementation outcomes, and increasing regulatory and community acceptance are likely to play a critical role in fostering blockchain adoption. We identify at least two ways to reduce uncertainty: (1) in-depth investigations of different use cases: as Risius and Spohrer (2017) also point out the number of business case studies is still very limited; (2) a suitability framework for blockchain applications: not all applications or all organisations may benefit from blockchain adoption, therefore a suitability framework like the one proposed by Misra and Mondal (2011) for cloud computing applications may represent a useful "reality check" for organisations. Building legitimacy around blockchain is extremely important in this context in order to increase regulatory and community (i.e. customers, employees, investors, and other stakeholders) acceptance (Rosati et al. 2016; Lynn et al. 2018). While Lynn et al. (2018) offer a first approach to this matter, further research is needed to gain a deep understanding of how blockchain legitimacy is changing over time, and if and how organisations are proactively trying to build it across multiple audiences (i.e. stakeholders). Regulators indeed have mostly taken a "wait and see" position on blockchain so far but they would be more likely to incentivise the adoption of a technology that is welcomed by multiple stakeholders. Finally, creating an analytical framework for measuring the value generated by blockchain investments may be extremely useful for building business cases for blockchain adoption. This is particularly relevant for financial services where financial resources are usually available but are allocated based on the return they are expected to generate. In this context being able to assess and quantify, with reasonable certainty, the impact of blockchain technology over the short and long term likely facilitates management buy-in.

This chapter has provided an overview of potential blockchain applications for the financial services sector and discussed related academic literature. As blockchain has the potential to automate many financial operations, it can generate significant gains in efficiency across the entire sector. For some intermediaries like brokers, clearing, and settlement houses though, those efficiency gains will result in lower revenues. The advent of blockchain technology poses significant challenges for these actors, who must radically reconsider their value propositions in order to stay in business. However, blockchain still remains at an early stage of development and the extent of changes it can generate in the financial sector is contingent to overcoming its current technical limitations and to increasing the acceptance of different stakeholders. A more collaborative approach to research across different academic disciplines and industry could be particularly fruitful in this context to advance the technology and realise its full potential.

References

Abdou, H. A., & Pointon, J. (2011). Credit scoring, statistical techniques and evaluation criteria: A review of the literature. *Intelligent Systems in Accounting, Finance and Management, 18*(2–3), 59–88.

Adhami, S., Giudici, G., & Martinazzi, S. (2018). Why do businesses go crypto? An empirical analysis of initial coin offerings. *Journal of Economics and Business*, forthcoming.

AFME. (2015). *Post trade explained—The role of post-trade services in the financial sector.* Available at: https://www.afme.eu/globalassets/downloads/publications/afme-post-trade-explained.pdf. Last accessed 10 August 2018.

Bank for International Settlements (BIS). (2017). *BIS international banking statistics at end-March 2017.* Available at: https://www.bis.org/statistics/rppb1707.pdf. Last accessed 15 August 2018.

Barth, M. E., Landsman, W. R., & Lang, M. H. (2008). International accounting standards and accounting quality. *Journal of Accounting Research, 46*(3), 467–498.

Batiz-Lazo, B. (2009). Emergence and evolution of ATM networks in the UK, 1967–2000. *Business History, 51*(1), 1–27.

Beck, R., Avital, M., Rossi, M., & Thatcher, J. B. (2017). Blockchain technology in business and information systems research. *Business & Information Systems Engineering, 59*(6), 381–384.

Beck, T., & Demirguc-Kunt, A. (2006). Small and medium-size enterprises: Access to finance as a growth constraint. *Journal of Banking and Finance, 30*(11), 2931–2943.

Beck, R., Czepluch, J. S., Lollike, N., & Malone, S. O. (2016). Blockchain—The gateway to trust-free cryptographic transactions. *24th European Conference on Information Systems (ECIS 2016)*. Istanbul, Turkey.

Beck, R., & Müller-Bloch, C. (2017). Blockchain as radical innovation: A framework for engaging with distributed ledgers. In *Proceedings of the 50th Hawaii International Conference on System Sciences*. Waikoloa Village.

Beck, R., Müller-Bloch, C., & King, J. L. (2018). Governance in the blockchain economy: A framework and research agenda. *Journal of the Association for Information Systems*.

Benos, E., Garratt, R., & Gurrola-Perez, P. (2017). The economics of distributed ledger technology for securities settlement. Available at SSRN: https://ssrn.com/abstract=3023779.

Bitcoin.org. (2018). *51% attack, majority hash rate attack*. Available at: https://bitcoin.org/en/glossary/51-percent-attack. Last accessed 29 August 2018.

Black, B. (2010). How to improve retail investor protection after the Dodd-Frank Wall Street Reform and Consumer Protection Act. *University of Pennsylvania Journal of Business Law, 13*, 59–106.

Böhme, R., Christin, N., Edelman, B., & Moore, T. (2015). Bitcoin: Economics, technology, and governance. *Journal of Economic Perspectives, 29*(2), 213–238.

Bott, J., & Milkau, U. (2017). Central bank money and blockchain: A payments perspective. *Journal of Payments Strategy & Systems, 11*(2), 145–157.

Broadridge. (2015). *Charting a path to a post-trade utility. How mutualized trade processing can reduce costs and help rebuild global bank ROE*. Broadridge White Paper. Available at: http://www.broadridge.com/broadridge-insights/Charting-a-Path-to-a-Post-Trade-Utility.html. Last accessed 10 August 2018.

Buitenhek, M. (2016). Understanding and applying blockchain technology in banking: Evolution or revolution? *Journal of Digital Banking, 1*(2), 111–119.

Byström, H. (2016). *Blockchains, real-time accounting and the future of credit risk modeling*. Department of Economics, Lund University. Available at: https://project.nek.lu.se/publications/workpap/papers/wp16_4.pdf. Last accessed 16 August 2018.

Capgemini Consulting. (2016). *Smart contracts in financial services: Getting from hype to reality*. Available at: https://www.capgemini.com/consulting/resources/blockchain-smart-contracts/. Last accessed 9 August 2018.

Catalini, C., & Gans, J. S. (2018). *Initial coin offerings and the value of crypto tokens* (No. w24418). National Bureau of Economic Research.

Chen, Y. (2018). Blockchain tokens and the potential democratization of entrepreneurship and innovation. *Business Horizons, 61*(4), 567–575.

Chiu, J., & Koeppl, T. V. (2018). Blockchain-based settlement for asset trading. Available at SSRN: https://ssrn.com/abstract=3203917.

Coyne, J. G., & McMickle, P. L. (2017). Can blockchains serve an accounting purpose? *Journal of Emerging Technologies in Accounting, 14*(2), 101–111.

Dai, J., & Vasarhelyi, M. A. (2017). Toward blockchain-based accounting and assurance. *Journal of Information Systems, 31*(3), 5–21.

Das, P., Verburg, R., Verbraeck, A., & Bonebakker, L. (2018). Barriers to innovation within large financial services firms: An in-depth study into disruptive and radical innovation projects at a bank. *European Journal of Innovation Management, 21*(1), 96–112.

Desai, F. (2015). *The evolution of Fintech.* Forbes. Available at: https://www.forbes.com/sites/falgunidesai/2015/12/13/the-evolution-of-fintech/#343661ff7175. Last accessed 9 August 2018.

European Central Bank. (2017). *Payment systems: Liquidity saving mechanisms in a distributed ledger environment.* Available at: https://www.ecb.europa.eu/pub/pdf/other/ecb.stella_project_report_september_2017.pdf. Last accessed 29 August 2018.

Fink, M. (2017). *Blockchains and data protections in the European Union* (Max Planck Institute for Innovation and Competition Research Paper 18(01)). Available at SSRN: https://ssrn.com/abstract=3080322.

Gardner, J. A. (2011). *Innovation and the future proof bank: A practical guide to doing different business-as-usual.* Chichester: Wiley.

Glaser, F. (2017). *Pervasive decentralisation of digital infrastructures: A framework for blockchain enabled system and use case analysis.* 50th Hawaii International Conference on System Sciences (HICSS 2017), Waikoloa, HI, USA.

Gomber, P., Kauffman, R. J., Parker, C., & Weber, B. W. (2018). On the fintech revolution: Interpreting the forces of innovation, disruption, and transformation in financial services. *Journal of Management Information Systems, 35*(1), 220–265.

Guo, Y., & Liang, C. (2016). Blockchain application and outlook in the banking industry. *Financial Innovation, 2*(1), 24.

Hawlitschek, F., Notheisen, B., & Teubner, T. (2018). The limits of trust-free systems: A literature review on blockchain technology and trust in the sharing economy. *Electronic Commerce Research and Applications, 29,* 50–63.

Higgins, S. (2018). *Method and system for net settlement by use of cryptographic promissory notes issued on a blockchain* (U.S. Patent Application No. 15/342). 463.

Howell, S. T., Niessner, M., & Yermack, D. (2018). *Initial coin offerings: Financing growth with cryptocurrency token sales* (No. w24774). National Bureau of Economic Research.

International Data Corporation (IDC). (2012). *IDC predictions 2012: Competing for 2020* (IDC #231720).

Jacobson, T., & Roszbach, K. (2003). Bank lending policy, credit scoring and value-at-risk. *Journal of Banking & Finance, 27*(4), 615–633.

Khapko, M., & Zoican, M. (2018). *Smart settlement. Society for Financial Studies (SFS) Cavalcade, 2017* (EFA 2017 Mannheim Meetings Paper; Rotman School of Management Working Paper No. 2881331; Swedish House of Finance Research Paper No. 17-4). Available at SSRN: https://ssrn.com/abstract=2881331.

Larios-Hernández, G. J. (2017). Blockchain entrepreneurship opportunity in the practices of the unbanked. *Business Horizons, 60*(6), 865–874.

Luftenegger, E., Angelov, S., van der Linden, E., & Grefen, P. (2010). *The state of the art of innovation-driven business models in the financial services industry* (Beta Report). Eindhoven University of Technology.

Lynn, T., Rosati, P., & Fox, G. (2018). *Legitimizing #Blockchain: An empirical analysis of firm level social media messaging on Twitter.* 26th European Conference on Information Systems (ECIS 2018).

Mackenzie, A. (2015). The fintech revolution. *London Business School Review, 26*(3), 50–53.

Mainelli, M., & Milne, A. (2016). *The impact and potential of blockchain on the securities transaction lifecycle* (SWIFT Institute Working Paper No. 2015-007).

Malinova, K., & Park, A. (2017). *Market design with blockchain technology.* Available at SSRN: https://ssrn.com/abstract=2785626.

McDowell, H. (2017). *Banks spent close to $100 billion on compliance last year.* Available at: https://www.thetradenews.com/banks-spent-close-to-100-billion-on-compliance-last-year/. Last accessed 10 August 2018.

McEvoy, M. J. (2014). *Enabling financial inclusion through "alternative data".* Bentonville, AR: Mastercard Advisors.

McKinsey & Co. (2016). *Global payments 2016: Strong fundamentals despite uncertain times.* Available at: https://www.mckinsey.com/~/media/McKinsey/Industries/Financial%20Services/Our%20Insights/A%20mixed%202015%20for%20the%20global%20payments%20industry/Global-Payments-2016.ashx. Last accessed 9 August 2018.

Mesropyan, E. (2016). *39% of the world's population does not have a bank account.* Available at: https://gomedici.com/39-of-the-worlds-population-does-not-have-a-bank-account/. Last accessed 9 August 2018.

Michel, S. (2014). Financial services and innovation. In T. Harrison & H. Estelami (Eds.), *The Routledge companion to financial services marketing* (p. 191). Abingdon: Routledge.

Micheler, E. (2015, November). Custody chains and asset values. Why crypto-securities are worth contemplating. *Cambridge Law Journal, 74*(3), 505–533.

Misra, S. C., & Mondal, A. (2011). Identification of a company's suitability for the adoption of cloud computing and modelling its corresponding Return on Investment. *Mathematical and Computer Modelling, 53*(3–4), 504–521.

Mougayar, W. (2016). *The business blockchain: Promise, practice and application of the next internet technology*. New Jersey: Wiley.

Moyano, J. P., & Ross, O. (2017). KYC optimization using distributed ledger technology. *Business & Information Systems Engineering, 59*(6), 411–423.

Nakamoto, S. (2008). *Bitcoin: A peer-to-peer electronic cash system*. Available at: https://bitcoin.org/bitcoin.pdf

Park, Y. (2016). *The inefficiencies of cross-border payments: How current forces are shaping the future*. Available at: http://euro.ecom.cmu.edu/resources/elibrary/epay/crossborder.pdf. Last accessed 15 August 2018.

Peters, G. W., & Panayi, E. (2016). Understanding modern banking ledgers through blockchain technologies: Future of transaction processing and smart contracts on the internet of money. In P. Tasca, T. Aste, L. Pelizzon, & N. Perony (Eds.), *Banking beyond banks and money* (pp. 239–278). Cham: Springer.

Prorokowski, L. (2015). MiFID II compliance—Are we ready? *Journal of Financial Regulation and Compliance, 23*(2), 196–206.

Risius, M., & Spohrer, K. (2017). A blockchain research framework. *Business & Information Systems Engineering, 59*(6), 385–409.

Rohr, J., and Wright, A. (2017). *Blockchain-based token sales, initial coin offerings, and the democratization of public capital markets* (Cardozo Legal Studies Research Paper No. 527; University of Tennessee Legal Studies Research Paper No. 338). Available at SSRN: https://ssrn.com/abstract=3048104 or http://dx.doi.org/10.2139/ssrn.3048104.

Roman, D., & Gatti, S. (2016, August). Towards a reference architecture for trusted data marketplaces: The credit scoring perspective. In *Open and Big Data (OBD), International Conference on IEEE* (pp. 95–101).

Rosati, P., Nair, B., & Lynn, T. (2016). #Bitcoin vs #blockchain: An exploratory study on the Bitcoin and blockchain discourse on Twitter. In *Proceedings of the 7th European Business Research Conference*.

Rosati, P., & Paulsson, V. (2017). Development of accounting information systems over time. In M. Quinn & E. Strauss (Eds.), *The Routledge companion to accounting information systems* (pp. 13–23). Abingdon: Routledge.

Ruce, P. J. (2011). Anti-money laundering: The challenges of know your customer legislation for private bankers and the hidden benefits for relationship management (the bright side of knowing your customer). *Banking Law Journal, 128,* 548–564.

Santander. (2015). *The Fintech 2.0 paper: Rebooting financial services*. Available at: https://www.finextra.com/finextra-downloads/newsdocs/the%20fintech%202%200%20paper.pdf. Last accessed 9 August 2018.

Seebacher, S., & Schüritz, R. (2017). Blockchain technology as an enabler of service systems: A structured literature review. In *International Conference on Exploring Services Science* (pp. 12–23). Cham: Springer.

Sheridan, I. (2017). MiFID II in the context of financial technology and regulatory technology. *Capital Markets Law Journal, 12*(4), 417–427.

Spezzati, S. (2017). *Banks are spending $20 Billion on compliance tech ahead of MiFID.* Available at: https://www.bloomberg.com/news/articles/2017-11-29/banks-are-spending-20-billion-on-compliance-tech-as-mifid-looms. Last accessed 10 August 2018.

Stahl, F., Schomm, F., & Vossen, G. (2014). Data marketplaces: An emerging species. In H.-M. Haav, A. Kalja, & T. Robal (Eds.), *Databases and information systems VIII* (pp. 145–158). Amsterdam: IOS Press.

Sun, J., Yan, J., & Zhang, K. Z. (2016). Blockchain-based sharing services: What blockchain technology can contribute to smart cities. *Financial Innovation, 2*(1), 2–26.

Tapscott, D., and Tapscott, A. (2016). The impact of the blockchain goes beyond financial services. *Harvard Business Review.* Retrieved May 25, 2017, from https://hbr.org/2016/05/the-impact-of-the-blockchain-goes-beyond-financial-services.

Terrell, E. (2010). *History of the American and NASDAQ stock exchanges.* In Library of Congress–Business Reference Services.

Thomas, L., Crook, J., & Edelman, D. (2017). *Credit scoring and its applications* (Vol. 2). Philadelphia: SIAM.

Thomson Reuters. (2018). *Cost of compliance 2018.* Available at: https://risk.thomsonreuters.com/content/dam/openweb/documents/pdf/risk/report/cost-of-compliance-special-report-2018.pdf. Last accessed 10 August 2018.

Tschorsch, F., & Scheuermann, B. (2016). Bitcoin and beyond: A technical survey on decentralized digital currencies. *IEEE Communications Surveys & Tutorials, 18*(3), 2084–2123.

Yeoh, P. (2017). Regulatory issues in blockchain technology. *Journal of Financial Regulation and Compliance, 25*(2), 196–208.

Yli-Huumo, J., Ko, D., Choi, S., Park, S., & Smolander, K. (2016). Where is current research on blockchain technology?—A systematic review. *PLoS ONE, 11*(10), e0163477.

Van der Elst, C., & Lafarre, A. (2018, July 24). *Blockchain and smart contracting for the shareholder community* (European Corporate Governance Institute (ECGI)—Law Working Paper No. 412/2018). Available at SSRN: https://ssrn.com/abstract=3219146.

Wang, Y., & Kogan, A. (2018). Designing confidentiality-preserving blockchain-based transaction processing systems. *International Journal of Accounting Information Systems, 30*, 1–18.

Weber, I., Xu, X., Riveret, R., Governatori, G., Ponomarev, A., & Mendling, J. (2016). Untrusted business process monitoring and execution using blockchain. In *International Conference on Business Process Management* (pp. 329–347). Cham: Springer.

World Bank. (2017). *The cost of sending remittances: June 2017.* Available at: http://www.worldbank.org/en/news/infographic/2017/06/15/the-cost-of-sending-remittances-june-2017. Last accessed 9 August 2018.

World Economic Forum (WEF). (2013). *The role of financial services in society—A multistakeholder compact.* Available at: http://www3.weforum.org/docs/WEF_FS_RoleFinancialServicesSociety_Report_2013.pdf. Last accessed 9 August 2018.

World Economic Forum (WEF). (2016). *The role of financial services in society—Understanding the impact of technology-enabled innovation on financial stability.* Available at: http://www3.weforum.org/docs/WEF_FS_RoleFinancialServicesSociety_Stability_Tech_Recommendations_2016.pdf. Last accessed 9 August 2018.

Wright, A., & De Filippi, P. (2015). *Decentralized blockchain technology and the rise of lex cryptographia.* Available at SSRN: 2580664.

Index

© The Editor(s) (if applicable) and The Author(s) 2019
T. Lynn et al. (eds.), *Disrupting Finance*, Palgrave Studies in Digital Business & Enabling Technologies, https://doi.org/10.1007/978-3-030-02330-0